T0156019

Lecture Notes in Computer Science 14321

Founding Editors

Gerhard Goos
Juris Hartmanis

The series Lecture Notes in Computer Science (LNCS), including its subseries Lecture Notes in Artificial Intelligence (LNAI) and Lecture Notes in Bioinformatics (LNBI), has established itself as a medium for the publication of new developments in computer science and information technology research, teaching, and education.

LNCS enjoys close cooperation with the computer science R & D community, the series counts many renowned academics among its volume editors and paper authors, and collaborates with prestigious societies. Its mission is to serve this international community by providing an invaluable service, mainly focused on the publication of conference and workshop proceedings and postproceedings. LNCS commenced publication in 1973.

Hossein Moosaei · Milan Hladík ·
Panos M. Pardalos

Editors

Dynamics of Information Systems

6th International Conference, DIS 2023
Prague, Czech Republic, September 3–6, 2023
Revised Selected Papers

 Springer

Editors
Hossein Moosaei ⓘ
Jan Evangelista Purkyně University
Ústí nad Labem-město, Czech Republic

Milan Hladík ⓘ
Charles University
Prague, Czech Republic

Panos M. Pardalos ⓘ
University of Florida
Gainesville, FL, USA

ISSN 0302-9743   ISSN 1611-3349 (electronic)
Lecture Notes in Computer Science
ISBN 978-3-031-50319-1   ISBN 978-3-031-50320-7 (eBook)
https://doi.org/10.1007/978-3-031-50320-7

This Springer imprint is published by the registered company Springer Nature Switzerland AG
The registered company address is: Gewerbestrasse 11, 6330 Cham, Switzerland

Paper in this product is recyclable.

Preface

The 6th International Conference on Dynamics of Information Systems (DIS 2023) was held from September 3 to 6, 2023, in Prague, Czech Republic, at the Department of Applied Mathematics at Charles University. Researchers and experts from more than 20 countries around the world participated in the event. This conference built on the success of previous DIS gatherings in the USA and focused on exploring the connections and unexplored areas in information science, optimization, operations research, machine learning, and artificial intelligence.

DIS 2023 brought together contributions from professionals in various fields, including information science, operations research, computer science, optimization, and electrical engineering. The discussions at DIS 2023 covered a wide range of topics from theoretical, algorithmic, and practical perspectives, offering readers valuable information, theories, and techniques.

The conference featured four keynote talks that shed light on important aspects of the conference themes. These talks were "Quantifying Model Uncertainty for Semantic Segmentation using RKHS Operators" by Jose C. Principe from the University of Florida, USA, "Towards a Dynamic Value of Information Theory" by Roman Belavkin from Middlesex University London, UK, "Diffusion Capacity of Single and Interconnected Networks" by Panos M. Pardalos from the University of Florida, USA, and "Search in Imperfect Information Games" by Martin Schmid from Google DeepMind, Canada.

We extend our gratitude to the authors for their valuable contributions and to the reviewers for their dedicated efforts in maintaining the high quality of the conference. This volume comprises 18 carefully selected and reviewed papers from 43 submissions, using a single-blind peer review process, with a minimum of two reviews per paper.

The editors express their gratitude to the organizers and sponsors of the DIS 2023 international conference:

- Charles University, Czech Republic
- Jan Evangelista Purkyně University, Czech Republic
- University of Florida, USA

We hope that this collection stands as a testament to the liveliness and rigor of the discussions at DIS 2023, fostering an atmosphere of curiosity and innovation that will shape the future of research and practice in information systems.

October 2023   Hossein Moosaei
                                                                  Milan Hladík
                                                              Panos M. Pardalos

Organization

General Chairs

Panos M. Pardalos   University of Florida, USA
Hossein Moosaei   Jan Evangelista Purkyně University, Czech
         Republic
Milan Hladík   Charles University, Czech Republic

Program Committee Chairs

Hossein Moosaei   Jan Evangelista Purkyně University, Czech
         Republic
Milan Hladík   Charles University, Czech Republic
Panos M. Pardalos   University of Florida, USA

Local Organizing Committee

Milan Hladík   Charles University, Czech Republic
Hossein Moosaei   Jan Evangelista Purkyně University, Czech
         Republic
David Hartman   Charles University, Czech Republic
Petr Kubera   Jan Evangelista Purkyně University, Czech
         Republic
Jiří Škvor   Jan Evangelista Purkyně University, Czech
         Republic
Martin Černý   Charles University, Czech Republic
Elif Garajová   Charles University, Czech Republic
Matyáš Lorenc   Charles University, Czech Republic
Jaroslav Horáček   Charles University, Czech Republic
Petra Příhodová   Charles University, Czech Republic

Program Committee

Pierre-Antoine Absil   University of Louvain, Belgium
Paula Amaral   Universidade Nova de Lisboa, Portugal

Contents

Using Data Mining Techniques to Analyze Facial Expression Motion Vectors

Mohamad Roshanzamir[1], Roohallah Alizadehsani[2(✉)], Mahdi Roshanzamir[3],
Afshin Shoeibi[4], Juan M. Gorriz[5], Abbas Khosravi[2], Saeid Nahavandi[6],
and U. Rajendra Acharya[7]

[1] Department of Computer Engineering, Faculty of Engineering, Fasa University,
74617-81189 Fasa, Iran
roshanzamir@fasau.ac.ir

[2] Institute for Intelligent Systems Research and Innovation (IISRI), Deakin University, Geelong,
VIC, Australia
{r.alizadehsani,abbas.khosravi}@deakin.edu.au

[3] Department of Electrical and Computer Engineering, University of Tabriz, Tabriz, Iran
roshanzamir@tabrizu.ac.ir

[4] Data Science and Computational Intelligence Institute, University of Granada, Granada, Spain

[5] Department of Signal Theory, Networking and Communications, Universidad de Granada,
Granada, Spain
gorriz@ugr.es

[6] Swinburne University of Technology, Hawthorn, VIC 3122, Australia
saeid.nahavandi@ieee.org

[7] School of Mathematics, Physics and Computing, University of Southern Queensland,
Springfield, Australia
Rajendra.Acharya@usq.edu.au

Abstract. Automatic recognition of facial expressions is a common problem in human-computer interaction. While humans can recognize facial expressions very easily, machines cannot do it as easily as humans. Analyzing facial changes during facial expressions is one of the methods used for this purpose by the machines. In this research, facial deformation caused by facial expressions is considered for automatic facial expression recognition by machines. To achieve this goal, the motion vectors of facial deformations are captured during facial expression using an optical flow algorithm. These motion vectors are then used to analyze facial expressions using some data mining algorithms. This analysis not only determined how changes in the face occur during facial expressions but can also be used for facial expression recognition. The facial expressions investigated in this research are happiness, sadness, surprise, fear, anger, and disgust. According to our research, these facial expressions were classified into 12 classes of facial motion vectors. We applied our proposed analysis mechanism to the extended Cohen-Kanade facial expression dataset. Our developed automatic facial expression system achieved 95.3%, 92.8%, and 90.2% accuracy using Deep Learning (DL), Support Vector Machine (SVM), and C5.0 classifiers, respectively. In addition, based on this research, it was determined which parts of the face have a greater impact on facial expression recognition.

S. Nahavandi—Associate Deputy Vice-Chancellor Research.

H. Moosaei et al. (Eds.): DIS 2023, LNCS 14321, pp. 1–19, 2024.
https://doi.org/10.1007/978-3-031-50320-7_1

Keywords: Automatic Facial Expression Recognition · Facial Expression Analysis · Optical Flow · Machine Learning · Data Mining · Emotion Classification · Feature Extraction · Human-Computer Interaction

1 Introduction

Facial expressions play an important role in human communication. According to Mehrabian research in [1], emotions appear 7% in spoken words, 38% in vocal utterances, and 55% in facial expressions. It is mentioned in [2] that non-verbal communication is more important than words in human communication. Nowadays, humans communicate not only with each other but also with machines. So, it would be ideal if machines could understand emotions. Automatic facial expression recognition can contribute to this goal. Research areas such as image processing, computer vision, machine learning, and data mining are used to achieve this goal. The results of this type of studies can be used in many applications, such as human-computer interaction (HCI). This is probably the most important application in automatic facial expression recognition [3]. Some other fields such as making animations and psychological research can also use the results of this type of research [4].

Although it is extremely easy for humans to recognize facial expressions, a machine that can do it like a human has not yet been built. Perfect automatic facial recognition systems must be fully automatic, independent of humans, and robust to all environmental conditions. They usually consist of three-step processes. These processes include face recognition, facial feature extraction, and facial expression classification [5]. Since these three processes are very broad, it is better to study them separately. Face recognition is not discussed in this paper and we assumed that face images are available. This work only focuses on facial feature extraction and facial expression classification. The novelties of this paper are:

1. In this research, we investigated and analyzed the motion vectors extracted from facial expression image sequences. These motion vectors are provided not only as the input to different data mining techniques to compare their performance but also to analyze the changes that occur in the face due to emotions.
2. So, a new motion vector dataset is created. This dataset contains motion vectors created because of the changes in the face due to facial expressions.
3. The area of the face that contributes to showing facial expression is analyzed. In addition to the predefined facial changes [6], our analysis leads to the discovery of several new types of changes.

The rest of this research was organized as follows. At first, the related works were explained in Sect. 2. The overview of the dataset used in this work was described in Sect. 3. Section 4 focused on the proposed method. Section 5 presents the experimental results. Finally, conclusions are given in Sect. 6.

2 Related Works

During the last three decades, many researchers focus their work on automatic facial expression recognition [7–17]. To this end, in some papers such as [18], different feature extraction techniques were combined with various classification algorithms to find the best combination that could be used for emotion recognition. Algorithms such as principal component analysis (PCA) [19, 20], stochastic neighbor embedding [21, 22], K-means clustering [23, 24], Haar feature selection technique (HFST) [25], distance vectors [26], feature distribution entropy [27, 28], gradient-based ternary texture patterns [29], hidden Markov models [30, 31], local binary patterns (LBP) [32, 33], dynamic Bayesian network [34], spectral regression [35], and conditional random field model [36] were also utilized for automatic facial expression recognition.

Other methods that nowadays are commonly used are artificial neural networks [37–39] and deep learning (DL)-based methods [40–45]. In [46], the Gabor wavelet transforms and convolutional neural network (CNN) [47–49] were combined to increase the accuracy of traditional recognition methods. In another research [50], three deep networks were designed to recognize emotions by synchronizing speech signals and image sequences. Fernandez et al. [51] proposed an end-to-end network and an attention model for automatic facial expression recognition. It focused on the human face and used a Gaussian space representation for expression recognition. In [52], a two-phase algorithm was proposed. A semi-supervised deep belief network approach was suggested to determine facial expressions. In addition, a gravitational search algorithm was applied to optimize some parameters in the proposed network to achieve a good accurate rate of classification for the automatic recognition of facial expressions. Finally, different surveys and reviews published about automatic facial expression recognition methods were presented in [53–56].

Although different methods were widely used for automatic facial expression recognition, there is not much research analyzing changes in the face because of facial expressions using data mining techniques. In this research, we cover this weakness of these types of research.

3 Dataset

In this study, facial expression image sequences downloaded from the Cohen-Kanade Expanded (CK+) facial expression dataset [57] were used as the test bed dataset. It is one of the most comprehensive datasets collected and processed in laboratory conditions. It has 593 series of facial images of 123 persons in the age range of 18 to 50. The size of images is 640 × 480 or 640 × 490 pixels. The sequence of each image starts in a neutral state and ends at the peak of one of the six main emotions happiness, sadness, surprise, fear, anger, and disgust [58].

4   Data Mining Techniques for Analyzing Motion Vectors of Facial Expression

Data mining is the automated process of discovering knowledge from data [59–63]. Indeed, it is a subject related to practical learning. The term *"learning"* refers to how data mining methods learn from data to improve their performance. Learning methods can map data based on that learning to a decision model to generate predictive results from new data. This type of decision model is called a classifier [64–67]. Today, these methods are widely used in various fields [68–71].

In this section, the method used for data collection was first introduced in Subsects. 4.1 and 4.2. Then, data cleaning and classification are explained in Subsect. 4.3.

4.1   Motion Vector Extraction

Facial expression causes temporary changes in facial features due to the movement of facial muscles. These changes last only a few seconds. Motion vectors representing facial deformations can be extracted from facial expression image sequences using a motion vector extraction algorithm. In this study, the optical flow algorithm [72] was applied to a facial expression image sequence to extract motion vectors. The optical flow algorithm is based on tracking points in multiple frames. Image preprocessing is done by converting images into grayscale. Then, the extra parts of the image were cut leading to an image size of about 280×330 pixels. Figure 1(a) and Fig. 1(b) show two sequences of preprocessed images representing disgust and happiness sequences, respectively.

There are different versions of the optical flow algorithm. The modified version proposed by Gautama-VanHull [73] was used in this study because it is one of the most accurate versions. Two examples of applying this algorithm to a sequence of facial expression images are shown in Fig. 1(c) and Fig. 1(d).

4.2   Face Segmentation

In this study, the face was divided into six areas shown in Fig. 2. Similar segmentation was performed in [10]. This segmentation is done manually or automatically specifying the eye and mouth positions in the first image of the input image sequence. In this research, the initial position is specified manually. However, it can be done automatically with high accuracy [74–77]. As it is clear in Fig. 2, axis number 1 connects two pupils. Axis #3 is perpendicular to axis #1 and divides it into two equal parts.

Figures 3 and 4 show why this division is done. Figure 3 shows the facial deformation according to Bassili's descriptions [6], while Fig. 4 shows other deformations experimentally extracted from image sequences in the CK+ dataset. The areas shown in Fig. 2 are then divided into smaller subsections as shown in Fig. 5(a). Dividing these regions into smaller parts allows for better analyzing the deformation of the face. Figure 5(b) shows nine axes pointing in different directions. The number and size of the subsections are the same in directions X1 and X4, X2 and X5, and X3 and X6, but their number and size are not considered to be fixed. In each subsection, the ratio of the number of vectors and their average length in the X and Y directions is calculated and used as raw data for data mining algorithms. So, if the number of these subsections is n, 3n features are generated for each image sequence.

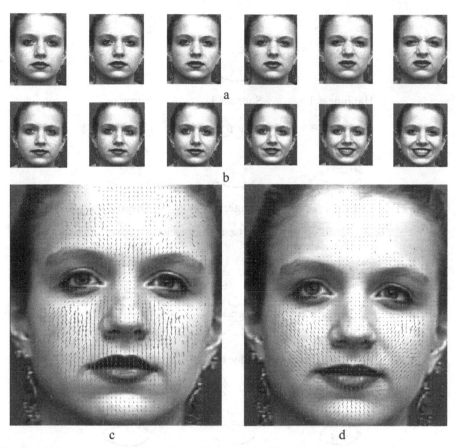

Fig. 1. Motion vector extraction using optical flow algorithm. (a) Image sequence of disgust (b) Image sequence of happiness. (c) Motion vectors extracted from a disgust image sequence and (d) Motion vectors extracted from the happiness image sequence.

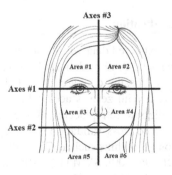

Fig. 2. Face segmentation into six areas

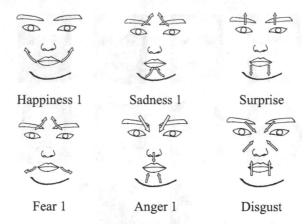

Fig. 3. Bassili's description of facial deformation in each emotion

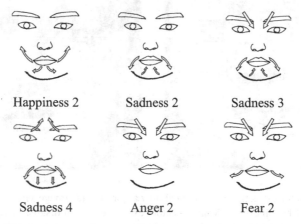

Fig. 4. Other types of face deformation in different emotions

4.3   Facial Expression Classification

At this phase of research, the extracted data will be analyzed. As the first step, the raw data must be cleaned. Missing data, extreme data, and outliers must be handled. Missing data were replaced with zeros. Missing data are generated in the regions without any motion vectors. So, replacing it with zero is acceptable. Extreme data was removed from the dataset. Finally, Outliers were replaced with the nearest values which are not considered as an outlier. A 10-fold cross-validation is then applied to this data 50 times. Some of the best classification algorithms such as C5.0 [78–82], CRT [83–86], QUEST [87–90], CHAID [91–94], DL [95–100], SVM [101–105], and Discriminate [106–108] were used for classification of this dataset. The features of this dataset are calculated according to Eqs. (1), (2), and (3).

$$P_{ij} = \frac{N_{ij}}{N} \tag{1}$$

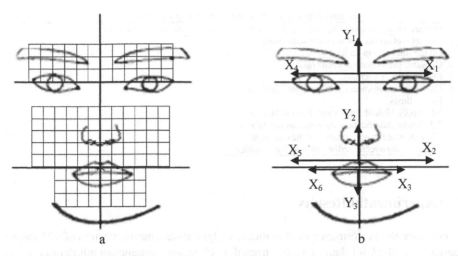

Fig. 5. a) Face areas are divided into smaller subsections. b) Nine different directions in which segmentations are done

$$LX_{ij} = \frac{1}{N_{ij}} \sum_{k=1}^{N_{ij}} lx_{ij}^k \tag{2}$$

$$LY_{ij} = \frac{1}{N_{ij}} \sum_{k=1}^{N_{ij}} ly_{ij}^k \tag{3}$$

where

- N is the total number of extracted motion vectors.
- N_{ij} is the number of extracted motion vectors in area i and subsection j.
- P_{ij} is the ratio of the number of motion vectors in area i and subsection j to the total number of extracted motion vectors.
- lx_{ij}^k is the length of motion vector k in direction X in section ij.
- ly_{ij}^k is the length of motion vector k in the direction Y in section ij.
- X_{ij} is the mean of motion vectors' length in direction X in area i and subsection j.
- LY_{ij} is the mean of motion vectors' length in the direction Y in area i and subsection j.

So, each feature vector includes the values of features calculated by Eqs. (1), (2), and (3). An example of a feature vector is illustrated in Fig. 6.

P_{11}	LX_{11}	LY_{11}	P_{12}	LX_{12}	LY_{12}	$\cdots$	P_{NM}	LX_{NM}	LY_{NM}

Fig. 6. A feature vector

The pseudo-code of the proposed method is shown in Algorithm 1.

Algorithm 1. Pseudo-code of the proposed method.

1. Prepare image sequence % Onset: natural state; Apex: one of six basic emotions.
2. Extract motion vectors in image sequences.
3. Divide images into six areas.
4. Generate subsections in each area.
5. Extract feature vectors
6. Manage missing values, extremes, and Outliers.
7. For 50 times
7.1. Apply 10-fold cross-validation on the dataset.
7.2. Apply classification algorithm on the dataset.
7.3. Estimate the performance of the classifier.
8. Calculate the overall performance of the classifier.

5 Experimental Results

To compare the performance of classification algorithms, motion vectors of 475 image sequences of the CK+ dataset were extracted. Each image sequence in this dataset represents one of the facial expressions shown in Figs. 3 and 4. Then, the features of extracted motion vectors were calculated by Eqs. 1, 2, and 3. Each image sequence was tested by applying different widths, heights, and numbers of subsections shown in Table 1. These adjustments have been tested to find the best adjustment. Their performance is reported in Table 2.

Table 1. The number of features in different adjustments

	#	Segments		Number of Segments in different directions						Total number of features
		Width*	Height*	X_1 & X_4	Y_1	X_2 & X_5	Y_2	X_3 & X_6	Y_3	
adjustment	1	5	5	3	3	4	3	2	3	162
	2	5	5	8	5	8	5	5	5	630
	3	5	15	2	1	3	1	2	1	42
	4	5	15	8	2	8	2	5	2	252
	5	5	20	5	2	5	2	2	1	132
	6	10	10	5	4	5	4	5	3	330
	7	10	10	5	5	5	5	5	5	450
	8	10	10	5	6	5	6	5	4	480
	9	10	15	5	3	5	3	3	2	216
	10	15	15	4	3	4	3	3	2	168
	11	15	15	5	5	5	5	5	5	450
	12	15	20	3	2	3	2	2	1	84

(*continued*)

Table 1. (*continued*)

#	Segments		Number of Segments in different directions						Total number of features
	Width[*]	Height[*]	X_1 & X_4	Y_1	X_2 & X_5	Y_2	X_3 & X_6	Y_3	
13	20	20	5	5	5	5	5	5	450
14	20	25	2	1	2	1	1	1	30
15	25	25	2	2	2	2	2	2	72
16	25	25	4	4	4	4	4	4	288
17	25	30	1	1	1	1	1	1	18
18	30	30	2	2	2	2	1	1	54
19	30	30	2	2	2	2	2	2	72
20	30	30	3	3	3	3	2	2	132
21	30	30	4	4	4	4	4	4	288
22	35	35	3	3	3	3	3	3	162
23	40	40	3	3	3	3	2	2	132
24	40	40	3	3	3	3	3	3	162
25	50	50	2	2	2	2	1	1	54

[*] In pixel

Table 2. Accuracy (%) of different classification algorithms in each adjustment. The best accuracy of each algorithm is bolded.

	#	Algorithms							
		DL	CRT	QUEST	CHAID	SVM	C5.0	Discriminant	*Overall AVG*
adjustment	1	81.4	52.9	52.9	58.2	75.5	88.1	63.9	*67.5*
	2	30.3	59.9	63.0	65.9	78.8	76.9	13.9	*55.5*
	3	75.7	52.1	51.5	53.3	77.4	89.1	65.2	*66.3*
	4	86.7	58.2	63.1	73.2	80.3	81.0	64.6	*72.4*
	5	83.3	61.9	58.9	69.0	77.7	87.2	71.3	*72.8*
	6	87.5	64.1	71.6	77.4	83.7	86.6	61.9	*76.1*
	7	90.4	69.5	71.2	74.9	82.4	69.5	58.3	*73.7*
	8	89.2	67.6	72.5	75.1	84.4	83.4	57.6	*75.7*
	9	88.8	65.3	72.7	76.9	87.9	73.7	63.3	*75.5*

(*continued*)

Table 2. (*continued*)

#	Algorithms							
	DL	CRT	QUEST	CHAID	SVM	C5.0	Discriminant	*Overall AVG*
10	91.5	67.9	76.0	75.7	87.5	78.6	63.6	*77.2*
11	91.4	67.9	69.3	**79.7**	87.3	65.5	64.3	*75.0*
12	86.4	61.7	67.5	68.2	76.2	84.7	84.0	*75.5*
13	89.7	67.3	74.3	76.9	85.4	64.7	61.6	*74.3*
14	84.4	56.6	68.9	65.8	81.3	88.3	76.9	*74.6*
15	90.0	64.5	70.5	72.4	85.4	84.3	82.2	*78.5*
16	91.3	69.7	**76.7**	72.1	86.6	72.1	67.8	*76.6*
17	81.2	61.9	61.6	62.9	88.3	86.4	76.4	*74.1*
18	89.1	69.1	72.3	71.2	82.4	87.6	81.7	*79.1*
19	92.8	**74.4**	71.6	77.3	**92.8**	**90.2**	81.2	***82.9***
20	88.6	69.8	70.6	75.7	85.9	78.5	71.4	*77.2*
21	**95.3**	67.9	70.6	74.6	83.2	73.4	67.5	*76.1*
22	88.3	64.3	74.9	76.7	82.9	72.1	67.9	*75.3*
23	92.7	68.9	71.8	72.7	87.8	76.6	77.2	*78.2*
24	91.4	68.9	70.1	78.9	86.7	70.6	69.1	*76.5*
25	87.8	66.9	70.4	74.0	81.0	89.1	**83.8**	*79.0*
Overall AVG	*85.8*	*64.8*	*68.6*	*71.9*	*83.6*	*79.9*	*67.9*	

According to the overall average of the examined algorithms shown in the last row of Table 2, DL, SVM, and C5.0 have the best performance in adjustment #21, adjustment #19, and adjustment #19 respectively with accuracy rates of 95.3%, 92.8%, and 90.2%. In adjustment #21, the width and height of the subsections are 30 pixels. Each area has four subsections in each direction. In this condition, 288 features were generated. In adjustment #19, the size of the subsections is 30 x 30 pixels. The number of subsections in each direction is two. So, there will be 72 features in our dataset.

Tables 3, 4, and 5 show the confusion matrix for DL, SVM, and C5.0 algorithms when different types of facial expression do not take into account. For example, both Happiness Type 1 and Happiness Type 2 are classified as the same class.

Table 3. Confusion matrix of DL algorithm in situation 21.

	Anger	Disgust	Fear	Happiness	Sadness	Surprise
Anger	89.9 ± 6.3	6.3 ± 2.0	1.0 ± 0.1	1.3 ± 0.7	1.0 ± 0.1	0.6 ± 0.5
Disgust	0.0 ± 0.0	96.7 ± 9.6	2.1 ± 0.7	1.2 ± 0.9	0.0 ± 0.0	0.0 ± 0.0
Fear	0.3 ± 0.1	0.3 ± 0.1	93.1 ± 7.8	4.8 ± 0.8	0.0 ± 0.0	1.5 ± 0.9
Happiness	0.0 ± 0.0	0.0 ± 0.0	1.2 ± 0.1	98.4 ± 8.6	0.0 ± 0.0	0.5 ± 0.0
Sadness	1.6 ± 0.1	0.2 ± 0.4	0.9 ± 0.1	0.2 ± 0.1	94.3 ± 5.6	2.7 ± 3.7
Surprise	0.0 ± 0.0	0.0 ± 0.0	0.7 ± 0.4	0.0 ± 0.0	0.2 ± 0.1	99.1 ± 7.3

Table 4. Confusion matrix of SVM algorithm in situation 19.

	Anger	Disgust	Fear	Happiness	Sadness	Surprise
Anger	90.5 ± 6.3	1.5 ± 0.5	1.0 ± 0.1	0.0 ± 0.0	7.0 ± 2.6	0.0 ± 0.0
Disgust	2.6 ± 1.1	93.9 ± 8.6	0.7 ± 0.1	1.4 ± 0.6	1.4 ± 0.6	0.0 ± 0.0
Fear	4.1 ± 0.9	0.0 ± 0.0	86.0 ± 3.9	7.1 ± 1.7	1.4 ± 0.7	1.4 ± 3.1
Happiness	0.0 ± 0.0	0.9 ± 1.4	2.8 ± 1.2	95.4 ± 6.9	0.9 ± 0.4	0.0 ± 0.0
Sadness	2.4 ± 0.0	1.1 ± 4.9	0.7 ± 0.4	1.1 ± 0.6	93.6 ± 8.4	1.1 ± 2.4
Surprise	0.0 ± 0.0	0.0 ± 0.0	1.2 ± 0.1	0.6 ± 0.3	0.6 ± 0.4	97.6 ± 7.4

Table 5. Confusion matrix of C5.0 algorithm in situation 19.

	Anger	Disgust	Fear	Happiness	Sadness	Surprise
Anger	89.8 ⊥ 8.9	2.2 ± 1.6	1.0 ± 0.8	0.0 ± 0.0	6.9 ± 0.8	0.0 ± 0.0
Disgust	1.2 ± 0.6	90.9 ± 4.7	0.7 ± 1.1	3.7 ± 0.6	1.2 ± 0.2	2.3 ± 0.4
Fear	0.7 ± 0.1	0.0 ± 0.0	86.1 ± 8.1	4.4 ± 1.3	1.3 ± 0.5	7.6 ± 4.4
Happiness	0.4 ± 0.2	0.7 ± 1.3	2.8 ± 0.2	92.7 ± 5.9	1.5 ± 0.7	1.9 ± 2.0
Sadness	3.2 ± 0.7	4.2 ± 7.5	0.7 ± 0.4	7.8 ± 1.2	83.1 ± 7.2	1.0 ± 3.2
Surprise	0.0 ± 0.0	0.0 ± 0.0	1.3 ± 0.2	0.3 ± 0.4	0.0 ± 0.0	98.4 ± 6.8

According to these confusion matrices, it is clear that misclassification usually occurs when there is a similarity between the position and direction of the motion vector. For example, only the direction of motion vectors in areas #2 and #3 can differentiate Disgust from anger Type 1, causing them to be misclassified when the direction of the motion vector is not properly recognized. Other high-rate misclassifications fall between Happiness and Fear. Because Happiness and Fear have the same motion vectors in the lower areas of the face, in about 1.2% of the tests, Fear is misclassified as Happy, as shown in Table 3. This is also the case in Tables 4 and 5. The highest rate of misclassifications in Tables 3, 4, and 5 are listed in Table 6.

Table 6. The highest rate of misclassifications is sorted according to the misclassification rate.

Algorithm name	Emotion	Misclassify as	Misclassification rate
C5.0	Sadness	Happiness	7.8
C5.0	Fear	Surprise	7.6
SVM	Fear	Happiness	7.1
SVM	Anger	Sadness	7.0
C5.0	Anger	Sadness	6.9
DL	Anger	Disgust	6.3
DL	Fear	Happiness	4.8

According to Tables 1 and 2, the three best adjustments for each algorithm are shown in Table 7. In the three algorithms CRT, SVM, and C5.0, adjustment #19 performed the best. This adjustment has the second rank in DL. Meanwhile, adjustment #19 has the best overall accuracy. Adjustment #19 is shown in Fig. 7. This is the best segmentation for facial expression recognition.

Table 7. Three best situations for each algorithm

Rank	DL	CRT	QUEST	CHAID	SVM	C5.0	Discriminant	Overall AVG
1	21	19	16	11	19	19	12	19
2	19	20	10	24	17	3	25	18
3	23	16	22	6	9	25	15	25

In Table 8, the features were arranged by the information gain algorithm [109] according to their importance for classification. In this table, subscripts i and j indicate the area number according to Fig. 2 and the subsection number according to Fig. 7(b). For example, LY_{14} indicates the length of the motion vector in the Y direction in area #1 and subsection #4. There are some interesting points in Table 8. For example, of the top 10 features, almost all of them are the length of motion vectors in the Y-direction. Of the 50 most important features, 48% are the length of the motion vector in the Y-direction. 42% are the length of the motion vectors in the X-direction. Only 10% of important features are the ratio of the number of motion vectors in different areas. It is clear that the length of the motion vector in the Y-direction is the most distinctive feature. In addition, according to the results shown in this table, area #3 and area #4 in Fig. 2 have less contribution than other areas in classification.

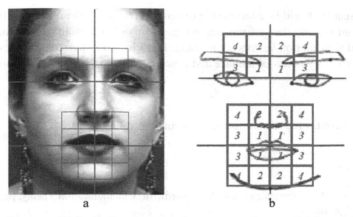

a   b

Fig. 7. a) adjustment #19 is the best segmentation to extract features b) The number of each subsection.

Table 8. The rank of features extracted by the Information Gain algorithm.

Rank	Feature	Rank	Feature	Rank	Feature	Rank	Feature	Rank	Feature	Rank	Feature
1	LY_{14}	11	LY_{54}	21	LY_{61}	31	LX_{14}	41	LX_{51}	51	LX_{12}
2	LY_{62}	12	LY_{64}	22	LY_{33}	32	LX_{62}	42	P_{51}	52	P_{41}
3	LY_{24}	13	LY_{63}	23	LX_{44}	33	LX_{61}	43	P_{61}	53	LX_{21}
4	LY_{22}	14	LX_{53}	24	LX_{54}	34	LX_{41}	44	P_{12}	54	P_{31}
5	LY_{52}	15	LX_{33}	25	LX_{34}	35	LY_{31}	45	LX_{13}	55	P_{54}
6	LY_{12}	16	LY_{53}	26	LY_{44}	36	LY_{41}	46	LX_{32}	56	P_{24}
7	LY_{21}	17	LY_{23}	27	LY_{34}	37	LX_{24}	47	P_{62}	57	P_{22}
8	LY_{11}	18	LY_{51}	28	LY_{32}	38	LX_{42}	48	LX_{22}	58	P_{14}
9	LY_{13}	19	LX_{63}	29	LX_{64}	39	LX_{52}	49	LX_{23}	59	LX_{11}
10	LX_{43}	20	LY_{43}	30	LY_{42}	40	LX_{31}	50	P_{52}	60	P_{33}

6   Conclusions and Future Work

Facial deformation analysis during an emotion is a critical subject for automatic facial expression recognition, human-computer interface design, and improving applications in security, medicine, psychology, entertainment, and education. This study not only analyzes the changes that occur due to emotions in the face but also tries to find better classification algorithms for the automatic recognition of facial expressions. This research is also looking for new face changes in addition to the default changes. To analyze the face deformation during facial expressions, a new data set containing motion vectors representing facial deformation during facial expressions was collected. Then, we tested which algorithm is better at classifying these deformations and which feature has the greatest impact for better classification. The data set generated in this search is new and there is no similar dataset. Therefore, no one does similar analyses in this field.

This research should be continued in different branches. Therefore, as future work, it is important to study the performance of algorithms when certain parts of the face are covered and their motion vectors are missing. It is also important to have these motion vectors from the side view of the face and conduct the same analysis from this point of view.

Conflict of Interest. The authors have no competing interests to declare.

References

1. Mehrabian, A.: Communication without words. In: Communication Theory, pp. 193–200. Routledge (2017)
2. Valstar, M.F., Mehu, M., Jiang, B., Pantic, M., Scherer, K.: Meta-analysis of the first facial expression recognition challenge. IEEE Trans. Syst. Man Cybern. Part B (Cybern.) **42**, 966–979 (2012)
3. Lisetti, C.L., Schiano, D.J.: Automatic facial expression interpretation: where human-computer interaction, artificial intelligence and cognitive science intersect. Pragmat. Cogn. **8**, 185–235 (2000)
4. Sultan Zia, M., Hussain, M., Arfan Jaffar, M.: A novel spontaneous facial expression recognition using dynamically weighted majority voting based ensemble classifier. Multimed. Tools Appl. **77**, 25537–25567 (2018)
5. Pantic, M., Rothkrantz, L.J.M.: Automatic analysis of facial expressions: the state of the art. IEEE Trans. Pattern Anal. Mach. Intell. **22**, 1424–1445 (2000)
6. Bassili, J.N.: Facial motion in the perception of faces and of emotional expression. J. Exp. Psychol. Hum. Percept. Perform. **4**, 373 (1978)
7. Vasanth, P., Nataraj, K.: Facial expression recognition using SVM classifier. Indon. J. Electr. Eng. Inform. (IJEEI) **3**, 16–20 (2015)
8. Abdulrahman, M., Eleyan, A.: Facial expression recognition using support vector machines. In: 2015 23nd Signal Processing and Communications Applications Conference (SIU), pp. 276–279. IEEE (2015)
9. Xu, X., Quan, C., Ren, F.: Facial expression recognition based on Gabor wavelet transform and histogram of oriented gradients. In: 2015 IEEE International Conference on Mechatronics and Automation (ICMA), pp. 2117–2122. IEEE (2015)
10. Naghsh-Nilchi, A.R., Roshanzamir, M.: An efficient algorithm for motion detection based facial expression recognition using optical flow. Proc. World Acad. Sci. Eng. Technol. **20**, 23–28 (2006)
11. Dhavalikar, A.S., Kulkarni, R.K.: Face detection and facial expression recognition system. In: 2014 International Conference on Electronics and Communication Systems (ICECS), pp. 1–7 (2014)
12. Roshanzamir, M., Naghsh Nilchi, A.R., Roshanzamir, M.: A new fuzzy rule-based approach for automatic facial expression recognition. 1st National Conference on Soft Computing. Undefined (1394)
13. Liliana, D.Y., Basaruddin, T., Widyanto, M.R., Oriza, I.I.D.: Fuzzy emotion: a natural approach to automatic facial expression recognition from psychological perspective using fuzzy system. Cogn. Process. **20**, 391–403 (2019)
14. Kirana, K.C., Wibawanto, S., Herwanto, H.W.: Facial emotion recognition based on viola-jones algorithm in the learning environment. In: 2018 International Seminar on Application for Technology of Information and Communication, pp. 406–410. IEEE (2018)

15. Happy, S., Routray, A.: Robust facial expression classification using shape and appearance features. In: 2015 Eighth International Conference on Advances in Pattern Recognition (ICAPR), pp. 1–5. IEEE (2015)
16. Chołoniewski, J., Chmiel, A., Sienkiewicz, J., Hołyst, J.A., Küster, D., Kappas, A.: Temporal Taylor's scaling of facial electromyography and electrodermal activity in the course of emotional stimulation. Chaos Solitons Fractals **90**, 91–100 (2016)
17. Tuncer, T., Dogan, S., Subasi, A.: A new fractal pattern feature generation function based emotion recognition method using EEG. Chaos, Solitons Fractals **144**, 110671 (2021)
18. Mehta, D., Siddiqui, M.F.H., Javaid, A.Y.: Recognition of emotion intensities using machine learning algorithms: a comparative study. Sensors **19**, 1897 (2019)
19. Varma, S., Shinde, M., Chavan, S.S.: Analysis of PCA and LDA features for facial expression recognition using SVM and hmm classifiers. In: Pawar, P.M., Ronge, B.P., Balasubramaniam, R., Vibhute, A.S., Apte, S.S. (eds.) Techno-Societal 2018, pp. 109–119. Springer, Cham (2020). https://doi.org/10.1007/978-3-030-16848-3_11
20. Islam, S.M.S., et al.: Machine learning approaches for predicting hypertension and its associated factors using population-level data from three south asian countries. Front. Cardiovasc. Med. **9**, 839379 (2022)
21. Lu, Y., Wang, S., Zhao, W., Zhao, Y.: WGAN-based robust occluded facial expression recognition. IEEE Access **7**, 93594–93610 (2019)
22. Nahavandi, D., Alizadehsani, R., Khosravi, A., Acharya, U.R.: Application of artificial intelligence in wearable devices: opportunities and challenges. Comput. Methods Programs Biomed. **213**, 106541 (2022)
23. Abdellatif, D., El Moutaouakil, K., Satori, K.: Clustering and Jarque-Bera normality test to face recognition. Procedia Comput. Sci. **127**, 246–255 (2018)
24. Malekzadeh, A., Zare, A., Yaghoobi, M., Alizadehsani, R.: Automatic diagnosis of epileptic seizures in EEG signals using fractal dimension features and convolutional autoencoder method. Big Data Cogn. Comput. **5**, 78 (2021)
25. Abdulrazaq, M.B., Mahmood, M.R., Zeebaree, S.R., Abdulwahab, M.H., Zebari, R.R., Sallow, A.B.: An analytical appraisal for supervised classifiers' performance on facial expression recognition based on relief-F feature selection. In: Journal of Physics: Conference Series, p. 012055. IOP Publishing (2021)
26. Barman, A., Dutta, P.: Facial expression recognition using distance and texture signature relevant features. Appl. Soft Comput. **77**, 88–105 (2019)
27. Wang, C., Wang, S., Liang, G.: Identity-and pose-robust facial expression recognition through adversarial feature learning. In: Proceedings of the 27th ACM International Conference on Multimedia, pp. 238–246 (2019)
28. Malekzadeh, A., Zare, A., Yaghoobi, M., Kobravi, H.-R., Alizadehsani, R.: Epileptic seizures detection in EEG signals using fusion handcrafted and deep learning features. Sensors **21**, 7710 (2021)
29. Saurav, S., Singh, S., Saini, R., Yadav, M.: Facial expression recognition using improved adaptive local ternary pattern. In: Chaudhuri, B.B., Nakagawa, M., Khanna, P., Kumar, S. (eds.) Proceedings of 3rd International Conference on Computer Vision and Image Processing. AISC, vol. 1024, pp. 39–52. Springer, Singapore (2020). https://doi.org/10.1007/978-981-32-9291-8_4
30. Rahul, M., Shukla, R., Goyal, P.K., Siddiqui, Z.A., Yadav, V.: Gabor filter and ICA-based facial expression recognition using two-layered hidden markov model. In: Gao, X.-Z., Tiwari, S., Trivedi, M.C., Mishra, K.K. (eds.) Advances in Computational Intelligence and Communication Technology. AISC, vol. 1086, pp. 511–518. Springer, Singapore (2021). https://doi.org/10.1007/978-981-15-1275-9_42

31. Rahul, M., Kohli, N., Agarwal, R.: Facial expression recognition using local multidirectional score pattern descriptor and modified hidden Markov model. Int. J. Adv. Intell. Paradig. **18**, 538–551 (2021)
32. Gogić, I., Manhart, M., Pandžić, I.S., Ahlberg, J.: Fast facial expression recognition using local binary features and shallow neural networks. Vis. Comput. **36**, 97–112 (2020)
33. Durmuşoğlu, A., Kahraman, Y.: Face expression recognition using a combination of local binary patterns and local phase quantization. In: 2021 International Conference on Communication, Control and Information Sciences (ICCISc), pp. 1–5. IEEE (2021)
34. Li, Y., Mavadati, S.M., Mahoor, M.H., Zhao, Y., Ji, Q.: Measuring the intensity of spontaneous facial action units with dynamic Bayesian network. Pattern Recogn. **48**, 3417–3427 (2015)
35. Wang, L., Wang, K., Li, R.: Unsupervised feature selection based on spectral regression from manifold learning for facial expression recognition. IET Comput. Vision **9**, 655–662 (2015)
36. Roshanzamir, M., et al.: Automatic facial expression recognition in an image sequence using conditional random field. In: 2022 IEEE 22nd International Symposium on Computational Intelligence and Informatics and 8th IEEE International Conference on Recent Achievements in Mechatronics, Automation, Computer Science and Robotics (CINTI-MACRo), pp. 000271–000278 (2022)
37. Alizadehsani, R., et al.: Diagnosis of coronary artery disease using data mining based on lab data and echo features. J. Med. Bioeng. **1** (2012)
38. Alizadehsani, R., et al.: Diagnosis of coronary arteries stenosis using data mining. J. Med. Signals Sens. **2**, 153 (2012)
39. Alizadehsani, R., Hosseini, M.J., Boghrati, R., Ghandeharioun, A., Khozeimeh, F., Sani, Z.A.: Exerting cost-sensitive and feature creation algorithms for coronary artery disease diagnosis. Int. J. Knowl. Discov. Bioinform. (IJKDB) **3**, 59–79 (2012)
40. Sharifrazi, D., et al.: CNN-KCL: Automatic myocarditis diagnosis using convolutional neural network combined with k-means clustering (2020)
41. Asgharnezhad, H., et al.: Objective evaluation of deep uncertainty predictions for COVID-19 detection. Sci. Rep. **12**, 815 (2022)
42. Roshanzamir, M., Alizadehsani, R., Roshanzamir, M., Shoeibi, A., Gorriz, J.M., Khosrave, A., Nahavandi, S.: What happens in Face during a facial expression? Using data mining techniques to analyze facial expression motion vectors. arXiv preprint arXiv:2109.05457 (2021)
43. Joloudari, J.H., et al.: DNN-GFE: a deep neural network model combined with global feature extractor for COVID-19 diagnosis based on CT scan images. EasyChair 2516–2314 (2021)
44. Ayoobi, N., et al.: Time series forecasting of new cases and new deaths rate for COVID-19 using deep learning methods. Results Phys. **27**, 104495 (2021)
45. Javan, A.A.K., et al.: Medical images encryption based on adaptive-robust multi-mode synchronization of Chen hyper-chaotic systems. Sensors **21**, 3925 (2021)
46. Qin, S., Zhu, Z., Zou, Y., Wang, X.: Facial expression recognition based on Gabor wavelet transform and 2-channel CNN. Int. J. Wavelets Multiresolut. Inf. Process. **18**, 2050003 (2020)
47. Shoushtarian, M., et al.: Objective measurement of tinnitus using functional near-infrared spectroscopy and machine learning. PLoS One **15**, e0241695 (2020)
48. Alizadehsani, R., et al.: Model uncertainty quantification for diagnosis of each main coronary artery stenosis. Soft. Comput. **24**, 10149–10160 (2020)
49. Zangooei, M.H., Habibi, J., Alizadehsani, R.: Disease diagnosis with a hybrid method SVR using NSGA-II. Neurocomputing **136**, 14–29 (2014)
50. Byun, S.-W., Lee, S.-P.: Human emotion recognition based on the weighted integration method using image sequences and acoustic features. Multimed. Tools Appl. 1–15 (2020)

51. Fernandez, P.D.M., Pena, F.A.G., Ren, T.I., Cunha, A.: FERAtt: facial expression recognition with attention net. arXiv preprint arXiv:1902.03284 3 (2019)
52. Alenazy, W.M., Alqahtani, A.S.: Gravitational search algorithm based optimized deep learning model with diverse set of features for facial expression recognition. J. Ambient. Intell. Humaniz. Comput. **12**, 1631–1646 (2021)
53. Alexandre, G.R., Soares, J.M., Thé, G.A.P.: Systematic review of 3D facial expression recognition methods. Pattern Recogn. **100**, 107108 (2020)
54. Li, S., Deng, W.: Deep facial expression recognition: a survey. IEEE Trans. Affect. Comput. **13**, 1195–1215 (2020)
55. Turan, C., Lam, K.-M.: Histogram-based local descriptors for facial expression recognition (FER): a comprehensive study. J. Vis. Commun. Image Represent. **55**, 331–341 (2018)
56. Abdullah, S.M.S., Abdulazeez, A.M.: Facial expression recognition based on deep learning convolution neural network: a review. J. Soft Comput. Data Min. **2**, 53–65 (2021)
57. Lucey, P., Cohn, J.F., Kanade, T., Saragih, J., Ambadar, Z., Matthews, I.: The extended Cohn-Kanade dataset (CK+): a complete dataset for action unit and emotion-specified expression. In: 2010 IEEE Computer Society Conference on Computer Vision and Pattern Recognition - Workshops, pp. 94–101 (2010)
58. The SAGE Encyclopedia of Theory in Psychology. SAGE Publications, Inc., Thousand Oaks (2016)
59. Tan, P.-N., Steinbach, M., Kumar, V.: Introduction to Data Mining. Pearson Education India (2016)
60. Moravvej, S.V., et al.: RLMD-PA: a reinforcement learning-based myocarditis diagnosis combined with a population-based algorithm for pretraining weights. Contrast Media Mol. Imaging **2022** (2022)
61. Kiss, N., et al.: Comparison of the prevalence of 21 GLIM phenotypic and etiologic criteria combinations and association with 30-day outcomes in people with cancer: a retrospective observational study. Clin. Nutr. **41**, 1102–1111 (2022)
62. Joloudari, J.H., et al.: Resource allocation optimization using artificial intelligence methods in various computing paradigms: a Review. arXiv preprint arXiv:2203.12315 (2022)
63. Alizadehsani, R., et al.: Factors associated with mortality in hospitalized cardiovascular disease patients infected with COVID-19. Immun. Inflamm. Dis. **10**, e561 (2022)
64. Bishop, C.M.: Pattern recognition. Mach. Learn. **128** (2006)
65. Alizadehsani, R., et al.: Hybrid genetic-discretized algorithm to handle data uncertainty in diagnosing stenosis of coronary arteries. Expert. Syst. **39**, e12573 (2022)
66. Khozeimeh, F., et al.: RF-CNN-F: random forest with convolutional neural network features for coronary artery disease diagnosis based on cardiac magnetic resonance. Sci. Rep. **12**, 11178 (2022)
67. Kakhi, K., Alizadehsani, R., Kabir, H.D., Khosravi, A., Nahavandi, S., Acharya, U.R.: The internet of medical things and artificial intelligence: trends, challenges, and opportunities. Biocybern. Biomed. Eng. **42**, 749–771 (2022)
68. Sharifrazi, D., et al.: Hypertrophic cardiomyopathy diagnosis based on cardiovascular magnetic resonance using deep learning techniques (2021)
69. Alizadehsani, R., et al.: Uncertainty-aware semi-supervised method using large unlabeled and limited labeled COVID-19 data. ACM Trans. Multimed. Comput. Commun. Appl. (TOMM) **17**, 1–24 (2021)
70. Shoeibi, A., et al.: Detection of epileptic seizures on EEG signals using ANFIS classifier, autoencoders and fuzzy entropies. Biomed. Signal Process. Control **73**, 103417 (2022)
71. Hassannataj Joloudari, J., et al.: GSVMA: a genetic support vector machine ANOVA method for CAD diagnosis. Front. Cardiovasc. Med. **8**, 760178 (2022)
72. Sonka, M., Hlavac, V., Boyle, R.: Image Processing, Analysis, and Machine Vision. Cengage Learning (2014)

73. Gautama, T., Hulle, M.A.V.: A phase-based approach to the estimation of the optical flow field using spatial filtering. IEEE Trans. Neural Netw. **13**, 1127–1136 (2002)
74. Li, W., Hua, Y., Liangzheng, X.: Mouth detection based on interest point. In: 2007 Chinese Control Conference, pp. 610–613 (2007)
75. Bao, P.T., Nguyen, H., Nhan, D.: A new approach to mouth detection using neural network. In: 2009 IITA International Conference on Control, Automation and Systems Engineering (case 2009), pp. 616–619 (2009)
76. Wang, Q., Yang, S., Li, X.W.: A fast mouth detection algorithm based on face organs. In: 2009 2nd International Conference on Power Electronics and Intelligent Transportation System (PEITS), pp. 250–252 (2009)
77. Asadifard, M., Shanbezadeh, J.: Automatic adaptive center of pupil detection using face detection and cdf analysis. In: Proceedings of the International Multiconference of Engineers and Computer Scientists, p. 3. Citeseer (2010)
78. Bujlow, T., Riaz, T., Pedersen, J.M.: A method for classification of network traffic based on C5.0 machine learning algorithm. In: 2012 International Conference on Computing, Networking and Communications (ICNC), pp. 237–241 (2012)
79. Nahavandi, S., et al.: A Comprehensive Review on Autonomous Navigation. arXiv preprint arXiv:2212.12808 (2022)
80. Eskandarian, R., et al.: Identification of clinical features associated with mortality in COVID-19 patients. Oper. Res. Forum **4**, 16 (2023)
81. Roshanzamir, M., et al.: Quantifying uncertainty in automated detection of Alzheimer's patients using deep neural network (2023)
82. Iqbal, M.S., Ahmad, W., Alizadehsani, R., Hussain, S., Rehman, R.: Breast cancer dataset, classification and detection using deep learning. In: Healthcare, p. 2395. MDPI (2022)
83. Blömer, J., Otto, M., Seifert, J.-P.: A new CRT-RSA algorithm secure against bellcore attacks. In: Proceedings of the 10th ACM Conference on Computer and Communications Security, pp. 311–320. Association for Computing Machinery, Washington D.C. (2003)
84. Nasab, R.Z., et al.: Deep Learning in spatially resolved transcriptomics: a comprehensive technical view. arXiv preprint arXiv:2210.04453 (2022)
85. Abedini, S.S., et al.: A critical review of the impact of candidate copy number variants on autism spectrum disorders. arXiv preprint arXiv:2303.03211 (2023)
86. Danaei, S., et al.: Myocarditis diagnosis: a method using mutual learning-based abc and reinforcement learning. In: 2022 IEEE 22nd International Symposium on Computational Intelligence and Informatics and 8th IEEE International Conference on Recent Achievements in Mechatronics, Automation, Computer Science and Robotics (CINTI-MACRo), pp. 000265–000270. IEEE (2022)
87. Bar-Itzhack, I.Y.: REQUEST-a recursive QUEST algorithm for sequential attitude determination. J. Guid. Control. Dyn. **19**, 1034–1038 (1996)
88. Joloudari, J.H., et al.: Application of artificial intelligence techniques for automated detection of myocardial infarction: a review. Physiol. Meas. (2022)
89. Kabir, H., et al.: Uncertainty aware neural network from similarity and sensitivity. arXiv preprint arXiv:2304.14925 (2023)
90. Khalili, H., et al.: Prognosis prediction in traumatic brain injury patients using machine learning algorithms. Sci. Rep. **13**, 960 (2023)
91. Lin, C.-L., Fan, C.-L.: Evaluation of CART, CHAID, and QUEST algorithms: a case study of construction defects in Taiwan. J. Asian Archit. Build. Eng. **18**, 539–553 (2019)
92. Nematollahi, M.A., et al.: Association and predictive capability of body composition and diabetes mellitus using artificial intelligence: a cohort study (2022)
93. Abbasi Habashi, S., Koyuncu, M., Alizadehsani, R.: A survey of COVID-19 diagnosis using routine blood tests with the aid of artificial intelligence techniques. Diagnostics **13**, 1749 (2023)

94. Karami, M., Alizadehsani, R., Argha, A., Dehzangi, I., Alinejad-Rokny, H.: Revolutionizing genomics with reinforcement learning techniques. arXiv preprint arXiv:2302.13268 (2023)
95. Khodatars, M., et al.: Deep learning for neuroimaging-based diagnosis and rehabilitation of autism spectrum disorder: a review. arXiv preprint arXiv:2007.01285 (2020)
96. Shoeibi, A., et al.: Applications of deep learning techniques for automated multiple sclerosis detection using magnetic resonance imaging: a review. arXiv preprint arXiv:2105.04881 (2021)
97. Mahamivanan, H., et al.: Material recognition for construction quality monitoring using deep learning methods. Constr. Innov. (2023)
98. Khozeimeh, F., et al.: ALEC: active learning with ensemble of classifiers for clinical diagnosis of coronary artery disease. Comput. Biol. Med. **158**, 106841 (2023)
99. Sadeghi, Z., et al.: A brief review of explainable artificial intelligence in healthcare. arXiv preprint arXiv:2304.01543 (2023)
100. Hong, L., et al.: GAN-LSTM-3D: An efficient method for lung tumour 3D reconstruction enhanced by attention-based LSTM. CAAI Trans. Intell. Technol. (2023)
101. Alizadehsani, R., et al.: A data mining approach for diagnosis of coronary artery disease. Comput. Methods Programs Biomed. **111**, 52–61 (2013)
102. Alizadehsani, R., et al.: Swarm intelligence in internet of medical things: A review. Sensors **23**, 1466 (2023)
103. Joloudari, J.H., et al.: BERT-deep CNN: state of the art for sentiment analysis of COVID-19 tweets. Soc. Netw. Anal. Min. **13**, 99 (2023)
104. Nahavandi, D., Alizadehsani, R., Khosravi, A.: Integration of machine learning with wearable technologies. Handb. Hum.-Mach. Syst. 383–396 (2023)
105. Kiss, N., et al.: Machine learning models to predict outcomes at 30-days using Global Leadership Initiative on Malnutrition combinations with and without muscle mass in people with cancer. J. Cachexia Sarcopenia Muscle (2023)
106. Park, C.H., Park, H.: A comparison of generalized linear discriminant analysis algorithms. Pattern Recogn. **41**, 1083–1097 (2008)
107. Nematollahi, M.A., et al.: Body composition predicts hypertension using machine learning methods: a cohort study. Sci. Rep. **13**, 6885 (2023)
108. Khozeimeh, F., et al.: Importance of wearable health monitoring systems using IoMT; Requirements, advantages, disadvantages and challenges. In: 2022 IEEE 22nd International Symposium on Computational Intelligence and Informatics and 8th IEEE International Conference on Recent Achievements in Mechatronics, Automation, Computer Science and Robotics (CINTI-MACRo), pp. 000245–000250. IEEE (2002)
109. Quinlan, J.R.: Induction of decision trees. Mach. Learn. **1**, 81–106 (1986)

A Numerical Scheme for a Generalized Fractional Derivative with Variable Order

Ricardo Almeida[✉][ID]

Center for Research and Development in Mathematics and Applications,
Department of Mathematics, University of Aveiro, Aveiro, Portugal
ricardo.almeida@ua.pt
https://www.ua.pt/pt/p/10321954

Abstract. The aim of this paper is to present an approximation formula for the Caputo fractional derivative of variable order, with dependence on an arbitrary kernel. For special cases of this kernel function, or the fractional order being constant, we recover some known formulas. This numerical method involves only integer-order derivatives, so any fractional problem can be approximated by an integer-order problem. Some numerical simulations end the paper, showing the effectiveness of our procedure.

Keywords: Fractional calculus · variable-order fractional derivative · approximation formula

1 Introduction

Problems of fractional order, and especially of variable order, are, in many situations, impossible to solve. Because of this, different numerical methods have been developed to overcome this difficulty. For example, using polynomial interpolation [13,14], different discretization formulas for the fractional operators [12,15,16], predictor-corrector algorithms [9,11], the Adomian decomposition method [10,21], or wavelets method [8]. In [6,7], Atanackovic and Stankovic presented a new method to deal with fractional derivatives. They showed that we can approximate them by an expansion formula involving only first order derivatives. This is a very useful procedure since, given any fractional problem, we can simply rewrite it as a first order problem and then apply any already known method from the classical literature. This study was followed by other works for different fractional operators (e.g. [3–5,17–20]).

We will consider fractional operators with variable fractional order, given by a function $\gamma_n : [a, b] \to (n - 1, n)$, where $n \in \mathbb{N}$. The kernel is given by a function $g : [a, b] \to \mathbb{R}$ of class C^n with $g'(t) > 0$, for all $t \in [a, b]$.

Work supported by Portuguese funds through the CIDMA - Center for Research and Development in Mathematics and Applications, and the Portuguese Foundation for Science and Technology (FCT-Fundação para a Ciência a a Tecnologia), within project UIDB/04106/2020.

Definition 1. *The generalized Riemann-Liouville fractional integrals of function u are defined by*

$$\mathbb{I}_{a+}^{\gamma_n} u(t) = \frac{1}{\Gamma(\gamma_n(t))} \int_a^t g'(s)(g(t) - g(s))^{\gamma_n(t)-1} u(s)\, ds, \quad \text{(left integral)}$$

and

$$\mathbb{I}_{b-}^{\gamma_n} u(t) = \frac{1}{\Gamma(\gamma_n(t))} \int_t^b g'(s)(g(s) - g(t))^{\gamma_n(t)-1} u(s)\, ds, \quad \text{(right integral)}.$$

Definition 2. *The generalized Caputo fractional derivatives of function u are defined by*

$$^C\mathbb{D}_{a+}^{\gamma_n} u(t) = \mathbb{I}_{a+}^{n-\gamma_n} \left(\frac{1}{g'(t)} \frac{d}{dt} \right)^n u(t), \quad \text{(left derivative)}$$

and

$$^C\mathbb{D}_{b-}^{\gamma_n} u(t) = \mathbb{I}_{b-}^{n-\gamma_n} \left(\frac{-1}{g'(t)} \frac{d}{dt} \right)^n u(t), \quad \text{(right derivative)}.$$

According to the definition

$$^C\mathbb{D}_{a+}^{\gamma_n} u(t) = \frac{1}{\Gamma(n - \gamma_n(t))} \int_a^t g'(s)(g(t) - g(s))^{n-1-\gamma_n(t)} \left(\frac{1}{g'(s)} \frac{d}{ds} \right)^n u(s)\, ds$$

and

$$^C\mathbb{D}_{b-}^{\gamma_n} u(t) = \frac{1}{\Gamma(n - \gamma_n(t))} \int_t^b g'(s)(g(s) - g(t))^{n-1-\gamma_n(t)} \left(\frac{-1}{g'(s)} \frac{d}{ds} \right)^n u(s)\, ds.$$

So, in contrast to the classical derivatives, these fractional derivatives are non-local operators, and, in the case of the left derivative, they contain the memory of the process from the start point. When the fractional order is constant, we recover the definition presented in [1]. In addition, if $g(t) = t$ we obtain the Caputo fractional derivatives, and for $g(t) = \ln t$, the Caputo–Hadamard fractional derivatives.

For example, for the power functions, their fractional derivatives are given by the formulae (cf. [2])

$$^C\mathbb{D}_{a+}^{\gamma_n} (g(t) - g(a))^\beta = \frac{\Gamma(\beta + 1)}{\Gamma(\beta - \overline{\gamma}_n(t) + 1)} (g(t) - g(a))^{\beta - \overline{\gamma}_n(t)}$$

and

$$^C\mathbb{D}_{b-}^{\gamma_n} (g(b) - g(t))^\beta = \frac{\Gamma(\beta + 1)}{\Gamma(\beta - \overline{\gamma}_n(t) + 1)} (g(b) - g(t))^{\beta - \overline{\gamma}_n(t)},$$

where $\beta > n - 1$.

For the next results, we recall the extension of the binomial formula for real numbers:

$$\binom{x}{k} (-1)^k = \frac{\Gamma(k - x)}{\Gamma(-x)k!},$$

for $x \in \mathbb{R}$ and $k \in \mathbb{N}$. In the next section, we present and prove our new approximation formulae for the generalized Caputo fractional derivatives of function u, using only integer order derivatives, and provide an upper bound formula for the error. In the end, we test the accuracy of the proposed method by comparing the exact fractional derivative of a function with its numerical approximation, and we also show how this procedure can be useful to solve fractional differential equations.

2   Theoretical Results

Theorem 1. *Consider a function* $u : [a,b] \to \mathbb{R}$ *of class* C^{n+m+1}, *where* $m \in \mathbb{N} \cup \{0\}$. *For* $N \in \mathbb{N}$ *with* $N \geq m+1$, *define the quantities*

$$A_k = 1 + \sum_{p=m-k+1}^{N} \frac{\Gamma(p-n+\gamma_n(t)-m)}{\Gamma(\gamma_n(t)-n-k)(p-m+k)!}, \quad k \in \{0,1,\ldots,m\},$$

$$B_k = \frac{\Gamma(k-n+\gamma_n(t)-m)}{\Gamma(n-\gamma_n(t))\Gamma(\gamma_n(t)-n+1)(k-m-1)!}, \quad k \in \{m+1,m+2,\ldots,N\},$$

and for $k \in \{m+1,m+2,\ldots,N\}$ *and for* $t \in [a,b]$, *define the function*

$$V_k(t) = \int_a^t g'(s)(g(s)-g(a))^{k-m-1}\left(\frac{1}{g'(s)}\frac{d}{ds}\right)^n u(s)ds.$$

Then,

$$^C\mathbb{D}_{a+}^{\gamma_n}u(t) = \sum_{k=0}^{m} A_k \frac{(g(t)-g(a))^{n-\gamma_n(t)+k}}{\Gamma(n-\gamma_n(t)+k+1)}\left(\frac{1}{g'(t)}\frac{d}{dt}\right)^{n+k} u(t)$$

$$+ \sum_{k=m+1}^{N} B_k(g(t)-g(a))^{n-\gamma_n(t)+m-k}V_k(t) + \Xi(N,t),$$

with

$$|\Xi(N,t)| \leq$$
$$\frac{(g(t)-g(a))^{n-\gamma_n(t)+m}}{\Gamma(n-\gamma_n(t)+m+1)}(t-a)M(t)\frac{\exp((n-\gamma_n(t)+m)^2+n-\gamma_n(t)+m)}{N^{n-\gamma_n(t)+m}(n-\gamma_n(t)+m)},$$

where

$$M(t) = \max_{s\in[a,t]}\left|g'(s)\left(\frac{1}{g'(s)}\frac{d}{ds}\right)^{n+m+1} u(s)\right|.$$

Proof. Performing an integration by parts, we get that

$$
{}^{C}\mathbb{D}_{a+}^{\gamma_n}u(t) = \left[\frac{1}{\Gamma(n-\gamma_n(t))} \cdot \frac{-(g(t)-g(s))^{n-\gamma_n(t)}}{n-\gamma_n(t)} \left(\frac{1}{g'(s)} \frac{d}{ds} \right)^n u(s) \right]_{s=a}^{s=t}
$$

$$
+ \frac{1}{\Gamma(n-\gamma_n(t))} \int_a^t \frac{(g(t)-g(s))^{n-\gamma_n(t)}}{n-\gamma_n(t)} \cdot \frac{d}{ds} \left(\frac{1}{g'(s)} \frac{d}{ds} \right)^n u(s)\, ds
$$

$$
= \frac{(g(t)-g(a))^{n-\gamma_n(t)}}{\Gamma(n-\gamma_n(t)+1)} \left(\frac{1}{g'(s)} \frac{d}{ds} \right)^n u(s) \Big|_{s=a}
$$

$$
+ \frac{1}{\Gamma(n-\gamma_n(t)+1)} \int_a^t g'(s)(g(t)-g(s))^{n-\gamma_n(t)} \left(\frac{1}{g'(s)} \frac{d}{ds} \right)^{n+1} u(s)\, ds.
$$

Doing another integration by parts, we get

$$
{}^{C}\mathbb{D}_{a+}^{\gamma_n}u(t) = \frac{(g(t)-g(a))^{n-\gamma_n(t)}}{\Gamma(n-\gamma_n(t)+1)} \left(\frac{1}{g'(s)} \frac{d}{ds} \right)^n u(s) \Big|_{s=a}
$$

$$
+ \left[\frac{1}{\Gamma(n-\gamma_n(t)+1)} \cdot \frac{-(g(t)-g(s))^{n-\gamma_n(t)+1}}{n-\gamma_n(t)+1} \left(\frac{1}{g'(s)} \frac{d}{ds} \right)^{n+1} u(s) \right]_{s=a}^{s=t}
$$

$$
+ \frac{1}{\Gamma(n-\gamma_n(t)+1)} \int_a^t \frac{(g(t)-g(s))^{n-\gamma_n(t)+1}}{n-\gamma_n(t)+1} \cdot \frac{d}{ds} \left(\frac{1}{g'(s)} \frac{d}{ds} \right)^{n+1} u(s)\, ds
$$

$$
= \sum_{k=0}^{1} \frac{(g(t)-g(a))^{n-\gamma_n(t)+k}}{\Gamma(n-\gamma_n(t)+k+1)} \left(\frac{1}{g'(s)} \frac{d}{ds} \right)^{n+k} u(s) \Big|_{s=a}
$$

$$
+ \frac{1}{\Gamma(n-\gamma_n(t)+2)} \int_a^t g'(s)(g(t)-g(s))^{n-\gamma_n(t)+1} \left(\frac{1}{g'(s)} \frac{d}{ds} \right)^{n+2} u(s)\, ds.
$$

If we perform $(m-1)$ more integration by parts, we arrive at

$$
{}^{C}\mathbb{D}_{a+}^{\gamma_n}u(t) = \sum_{k=0}^{m} \frac{(g(t)-g(a))^{n-\gamma_n(t)+k}}{\Gamma(n-\gamma_n(t)+k+1)} \left(\frac{1}{g'(s)} \frac{d}{ds} \right)^{n+k} u(s) \Big|_{s=a}
$$

$$
+ \frac{1}{\Gamma(n-\gamma_n(t)+m+1)} \int_a^t g'(s)(g(t)-g(s))^{n-\gamma_n(t)+m}
$$

$$
\times \left(\frac{1}{g'(s)} \frac{d}{ds} \right)^{n+m+1} u(s)\, ds.
$$

Using the following expansion formula

$$
(g(t)-g(s))^{n-\gamma_n(t)+m} = (g(t)-g(a))^{n-\gamma_n(t)+m} \left(1 - \frac{g(s)-g(a)}{g(t)-g(a)} \right)^{n-\gamma_n(t)+m}
$$

$$
= (g(t)-g(a))^{n-\gamma_n(t)+m} \sum_{k=0}^{\infty} \binom{n-\gamma_n(t)+m}{k} (-1)^k \left(\frac{g(s)-g(a)}{g(t)-g(a)} \right)^k
$$

$$
= (g(t)-g(a))^{n-\gamma_n(t)+m} \sum_{k=0}^{\infty} \frac{\Gamma(k-n+\gamma_n(t)-m)}{\Gamma(\gamma_n(t)-n-m)k!} \left(\frac{g(s)-g(a)}{g(t)-g(a)} \right)^k,
$$

we obtain the following decomposition formula

$$
{}^{C}\mathbb{D}_{a+}^{\gamma_n} u(t) = \sum_{k=0}^{m} \frac{(g(t) - g(a))^{n-\gamma_n(t)+k}}{\Gamma(n - \gamma_n(t) + k + 1)} \left(\frac{1}{g'(s)} \frac{d}{ds} \right)^{n+k} u(s) \Big|_{s=a}
$$

$$
+ \frac{(g(t) - g(a))^{n-\gamma_n(t)+m}}{\Gamma(n - \gamma_n(t) + m + 1)} \sum_{k=0}^{N} \frac{\Gamma(k - n + \gamma_n(t) - m)}{\Gamma(\gamma_n(t) - n - m)k!(g(t) - g(a))^k}
$$

$$
\times \int_{a}^{t} g'(s)(g(s) - g(a))^k \left(\frac{1}{g'(s)} \frac{d}{ds} \right)^{n+m+1} u(s) \, ds + \Xi(N, t),
$$

where

$$
\Xi(N, t) = \frac{(g(t) - g(a))^{n-\gamma_n(t)+m}}{\Gamma(n - \gamma_n(t) + m + 1)} \sum_{k=N+1}^{\infty} \frac{\Gamma(k - n + \gamma_n(t) - m)}{\Gamma(\gamma_n(t) - n - m)k!(g(t) - g(a))^k}
$$

$$
\times \int_{a}^{t} g'(s)(g(s) - g(a))^k \left(\frac{1}{g'(s)} \frac{d}{ds} \right)^{n+m+1} u(s) \, ds.
$$

With respect to the first sum, consider the case when $k = 0$ and, for the remaining terms ($k \in \{1, \ldots, N\}$), integrate by parts the integral and we get that the Caputo fractional derivative can be written as

$$
\sum_{k=0}^{m-1} \frac{(g(t) - g(a))^{n-\gamma_n(t)+k}}{\Gamma(n - \gamma_n(t) + k + 1)} \left(\frac{1}{g'(s)} \frac{d}{ds} \right)^{n+k} u(s) \Big|_{s=a}
$$

$$
+ \frac{(g(t) - g(a))^{n-\gamma_n(t)+m}}{\Gamma(n - \gamma_n(t) + m + 1)} \left(\frac{1}{g'(t)} \frac{d}{dt} \right)^{n+m} u(t)
$$

$$
+ \frac{(g(t) - g(a))^{n-\gamma_n(t)+m}}{\Gamma(n - \gamma_n(t) + m + 1)} \sum_{k=1}^{N} \frac{\Gamma(k - n + \gamma_n(t) - m)}{\Gamma(\gamma_n(t) - n - m)k!(g(t) - g(a))^k}
$$

$$
\times \left[\left[(g(s) - g(a))^k \left(\frac{1}{g'(s)} \frac{d}{ds} \right)^{n+m} u(s) \right]_{s=a}^{s=t} \right.
$$

$$
\left. - \int_{a}^{t} kg'(s)(g(s) - g(a))^{k-1} \left(\frac{1}{g'(s)} \frac{d}{ds} \right)^{n+m} u(s) \, ds \right] + \Xi(N, t)
$$

$$
= \sum_{k=1}^{m-1} \frac{(g(t) - g(a))^{n-\gamma_n(t)+k}}{\Gamma(n - \gamma_n(t) + k + 1)} \left(\frac{1}{g'(s)} \frac{d}{ds} \right)^{n+k} u(s) \Big|_{s=a}
$$

$$
+ \frac{(g(t) - g(a))^{n-\gamma_n(t)+m}}{\Gamma(n - \gamma_n(t) + m + 1)} \left(\frac{1}{g'(t)} \frac{d}{dt} \right)^{n+m} u(t) A_m
$$

$$
+ \frac{(g(t) - g(a))^{n-\gamma_n(t)+m-1}}{\Gamma(n - \gamma_n(t) + m)} \sum_{k=1}^{N} \frac{\Gamma(k - n + \gamma_n(t) - m)}{\Gamma(\gamma_n(t) - n - m + 1)(k - 1)!(g(t) - g(a))^{k-1}}
$$

$$
\times \int_{a}^{t} g'(s)(g(s) - g(a))^{k-1} \left(\frac{1}{g'(s)} \frac{d}{ds} \right)^{n+m} u(s) \, ds + \Xi(N, t)
$$

Now, consider in the last sum the case $k = 1$ and for the remaining terms, integrate them by parts and we obtain

$$\sum_{k=0}^{m-2} \frac{(g(t) - g(a))^{n-\gamma_n(t)+k}}{\Gamma(n - \gamma_n(t) + k + 1)} \left(\frac{1}{g'(s)}\frac{d}{ds}\right)^{n+k} u(s)\Big|_{s=a}$$

$$+ \frac{(g(t) - g(a))^{n-\gamma_n(t)+m}}{\Gamma(n - \gamma_n(t) + m + 1)} \left(\frac{1}{g'(t)}\frac{d}{dt}\right)^{n+m} u(t)A_m$$

$$+ \frac{(g(t) - g(a))^{n-\gamma_n(t)+m-1}}{\Gamma(n - \gamma_n(t) + m)} \left(\frac{1}{g'(t)}\frac{d}{dt}\right)^{n+m-1} u(t)$$

$$+ \frac{(g(t) - g(a))^{n-\gamma_n(t)+m-1}}{\Gamma(n - \gamma_n(t) + m)} \sum_{k=2}^{N} \frac{\Gamma(k - n + \gamma_n(t) - m)}{\Gamma(\gamma_n(t) - n - m + 1)(k - 1)!(g(t) - g(a))^{k-1}}$$

$$\times \left[\left[(g(s) - g(a))^{k-1} \left(\frac{1}{g'(s)}\frac{d}{ds}\right)^{n+m-1} u(s)\right]_{s=a}^{s=t}\right.$$

$$\left. - \int_a^t (k-1)g'(s)(g(s) - g(a))^{k-2}\left(\frac{1}{g'(s)}\frac{d}{ds}\right)^{n+m-1} u(s)\,ds\right]$$

$$+ \Xi(N, t)$$

$$= \sum_{k=0}^{m-2} \frac{(g(t) - g(a))^{n-\gamma_n(t)+k}}{\Gamma(n - \gamma_n(t) + k + 1)} \left(\frac{1}{g'(s)}\frac{d}{ds}\right)^{n+k} u(s)\Big|_{s=a}$$

$$+ \frac{(g(t) - g(a))^{n-\gamma_n(t)+m}}{\Gamma(n - \gamma_n(t) + m + 1)} \left(\frac{1}{g'(t)}\frac{d}{dt}\right)^{n+m} u(t)A_m$$

$$+ \frac{(g(t) - g(a))^{n-\gamma_n(t)+m-1}}{\Gamma(n - \gamma_n(t) + m)} \left(\frac{1}{g'(t)}\frac{d}{dt}\right)^{n+m-1} u(t)A_{m-1}$$

$$+ \frac{(g(t) - g(a))^{n-\gamma_n(t)+m-2}}{\Gamma(n - \gamma_n(t) + m - 1)} \sum_{k=2}^{N} \frac{\Gamma(k - n + \gamma_n(t) - m)}{\Gamma(\gamma_n(t) - n - m + 2)(k - 2)!(g(t) - g(a))^{k-2}}$$

$$\times \int_a^t g'(s)(g(s) - g(a))^{k-2}\left(\frac{1}{g'(s)}\frac{d}{ds}\right)^{n+m-1} u(s)\,ds + \Xi(N, t).$$

Repeating this procedure, we arrive at the desired formula. To end, we will obtain the upper bound for the error. For that, simply observe that

$$\left|\left(\frac{g(s) - g(a)}{g(t) - g(a)}\right)^k\right| \leq 1, \quad \forall s \in [a, t],$$

and that

$$\sum_{k=N+1}^{\infty} \left| \binom{n - \gamma_n(t) + m}{k} (-1)^k \right|$$

$$\leq \sum_{k=N+1}^{\infty} \frac{\exp((n - \gamma_n(t) + m)^2 + n - \gamma_n(t) + m)}{k^{n-\gamma_n(t)+m+1}}$$

$$\leq \int_N^{\infty} \frac{\exp((n - \gamma_n(t) + m)^2 + n - \gamma_n(t) + m)}{k^{n-\gamma_n(t)+m+1}} \, dk$$

$$= \frac{\exp((n - \gamma_n(t) + m)^2 + n - \gamma_n(t) + m)}{N^{n-\gamma_n(t)+m}(n - \gamma_n(t) + m)}.$$

Observe that $\Xi(N,t)$ goes to zero as $N \to \infty$, and so the approximation

$${}^C\mathbb{D}_{a+}^{\gamma_n} u(t) \approx \sum_{k=0}^{m} A_k \frac{(g(t) - g(a))^{n-\gamma_n(t)+k}}{\Gamma(n - \gamma_n(t) + k + 1)} \left(\frac{1}{g'(t)} \frac{d}{dt} \right)^{n+k} u(t)$$

$$+ \sum_{k=m+1}^{N} B_k (g(t) - g(a))^{n-\gamma_n(t)+m-k} V_k(t)$$

makes sense. A similar formula can be deduced for the right fractional derivative:

Theorem 2. *Consider a function* $u : [a,b] \to \mathbb{R}$ *of class* C^{n+m+1}, *where* $m \in \mathbb{N} \cup \{0\}$. *For* $N \in \mathbb{N}$ *with* $N \geq m + 1$, *define the quantities* A_k *(for* $k \in \{0, 1, \ldots, m\}$*) and* B_k *(for* $k \in \{m+1, m+2, \ldots, N\}$*) as defined in Theorem 1. For* $k \in \{m+1, m+2, \ldots, N\}$ *and for* $t \in [a,b]$, *define the function*

$$V_k(t) = \int_t^b g'(s)(g(b) - g(s))^{k-m-1} \left(-\frac{1}{g'(s)} \frac{d}{ds} \right)^n u(s) ds.$$

Then,

$${}^C\mathbb{D}_{b-}^{\gamma_n} u(t) = \sum_{k=0}^{m} A_k \frac{(g(b) - g(t))^{n-\gamma_n(t)+k}}{\Gamma(n - \gamma_n(t) + k + 1)} \left(-\frac{1}{g'(t)} \frac{d}{dt} \right)^{n+k} u(t)$$

$$+ \sum_{k=m+1}^{N} B_k (g(b) - g(t))^{n-\gamma_n(t)+m-k} V_k(t) + \Xi(N,t),$$

with

$$|\Xi(N,t)| \leq$$
$$\frac{(g(b) - g(t))^{n-\gamma_n(t)+m}}{\Gamma(n - \gamma_n(t) + m + 1)} (b - t) M(t) \frac{\exp((n - \gamma_n(t) + m)^2 + n - \gamma_n(t) + m)}{N^{n-\gamma_n(t)+m}(n - \gamma_n(t) + m)}.$$

where

$$M(t) = \max_{s \in [t,b]} \left| g'(s) \left(-\frac{1}{g'(s)} \frac{d}{ds} \right)^{n+m+1} u(s) \right|.$$

3   Examples

To test the accuracy of our results, consider the function $u(t) = (g(t) - g(a))^3$. In this case,

$$^C\mathbb{D}_{a+}^{\gamma_n} u(t) = \frac{3!}{\Gamma(4 - \gamma_n(t))}(g(t) - g(a))^{3-\gamma_n(t)}.$$

We fix the initial point $a = 0$, the fractional order as $\gamma_1(t) = (t+1)/4$, with $t \in [0, 1]$, and two values for m, as given in Theorem 1, are considered. We present a comparison between the fractional derivative of function u with its numerical approximation, as presented in Theorem 1, for two different kernels: $g(t) = \sin(t)$ (Fig. 1) and $g(t) = \exp(t)$ (Fig. 2), with $m = 0$, and the same study but for $m = 2$ (Figs. 3 and 4, respectively). As can be easily observed, by increasing the value of m we obtain better accuracy for the method. For our next example, we consider the fractional differential equation

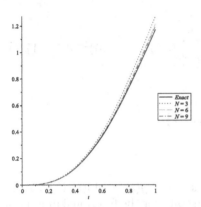

Fig. 1. $g(t) = \sin(t)$, $m = 0$

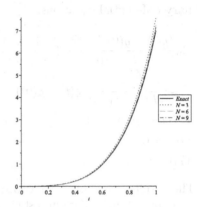

Fig. 2. $g(t) = \exp(t)$, $m = 0$

$$^C\mathbb{D}_{0+}^{\gamma_1} u(t) = \frac{3!}{\Gamma(4 - \gamma_1(t))}(g(t) - g(0))^{3-\gamma_1(t)}, \quad u(0) = 0,$$

whose obvious solution is $u(t) = (g(t) - g(0))^3$. Again, we consider $\gamma_1(t) = (t+1)/4$, with $t \in [0, 1]$, and $m = 0$. Considering the numerical approximation given in Theorem 1, we can approximate this fractional problem by a system of

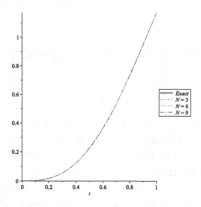

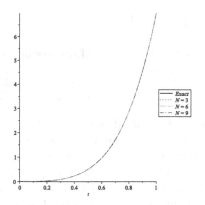

Fig. 3. $g(t) = \sin(t)$, $m = 2$ **Fig. 4.** $g(t) = \exp(t)$, $m = 2$

ordinary differential equations:

$$
\begin{cases}
A_0 \dfrac{(g(t) - g(0))^{1-\gamma_1(t)}}{\Gamma(2 - \gamma_1(t))} \left(\dfrac{1}{g'(t)} \dfrac{d}{dt} \right) u(t) + \sum_{k=1}^{N} B_k (g(t) - g(0))^{1-\gamma_1(t)-k} V_k(t) \\[2ex]
\quad = \dfrac{3!}{\Gamma(4 - \gamma_1(t))} (g(t) - g(0))^{3-\gamma_1(t)} \\[2ex]
V_k'(t) = (g(t) - g(0))^{k-1} u'(t), \quad k \in \{1, \dots, N\} \\[1ex]
u(0) = 0 \\[1ex]
V_k(0) = 0, \quad k \in \{1, \dots, N\}.
\end{cases}
$$

The results are exhibit for two different kernels: $g(t) = \sin(t)$ (Fig. 5) and $g(t) = \exp(t)$ (Fig. 6). For our last example, we consider the fractional differential equation

$$
{}^{C}\mathbb{D}_{0+}^{\gamma_1} u(t) = \frac{1}{u(t)}, \quad u(0) = 1.
$$

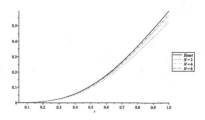

Fig. 5. $g(t) = \sin(t)$ **Fig. 6.** $g(t) = \exp(t)$

In this case, we do not know the exact solution to the problem. So, to test the accuracy of the method, we will compute the distance between the solutions for different values of N. The fractional order is $\gamma_1(t) = (t+1)/4$, with $t \in [0,1]$, and $m = 0$. So, the first order system of differential equations that we will consider, as an approximation for the fractional one, will be given by

$$
\begin{cases}
A_0 \dfrac{(g(t) - g(0))^{1-\gamma_1(t)}}{\Gamma(2 - \gamma_1(t))} \left(\dfrac{1}{g'(t)} \dfrac{d}{dt} \right) u(t) + \sum_{k=1}^{N} B_k (g(t) - g(0))^{1-\gamma_1(t)-k} V_k(t) \\
\qquad = \dfrac{1}{u(t)} \\
V_k'(t) = (g(t) - g(0))^{k-1} u'(t), \quad k \in \{1, \ldots, N\} \\
u(0) = 1 \\
V_k(0) = 0, \quad k \in \{1, \ldots, N\}.
\end{cases}
$$

In Figs. 7 and 8 we present the results, with respect to $g(t) = \sin(t)$ and $g(t) = \exp(t)$, respectively. For Case 1, we plot the graph of the absolute value of the difference between the solutions for the cases $N = 2$ and $N = 4$, and for Case 2, we consider $N = 4$ and $N = 6$. As can be observed, increasing the value of N makes the solutions closer, as expected.

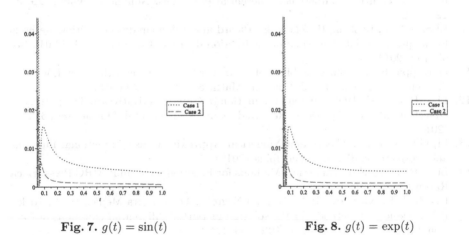

Fig. 7. $g(t) = \sin(t)$ **Fig. 8.** $g(t) = \exp(t)$

4 Conclusion

We presented approximation formulas for a certain class of fractional derivatives, with dependence on arbitrary kernels. These formulas only depend on integer-order derivatives of the initial function, and so the fractional model can be rewritten as a classical model. The order of convergence is $n - \gamma_n(t) + m$, so to have good accuracy, we need higher-order derivatives. In the future, it is important to develop other methods that do not require as much smoothness in the function.

References

1. Almeida, R.: A Caputo fractional derivative of a function with respect to another function. Commun. Nonlinear Sci. Numer. Simul. **44**, 460–481 (2017)
2. Almeida, R.: Variational problems of variable fractional order involving arbitrary kernels. AIMS Math. **7**(10), 18690–18707 (2022)
3. Almeida, R., Bastos, N.R.O.: A numerical method to solve higher-order fractional differential equations. Mediterr. J. Math. **13**, 1339–1352 (2016)
4. Atanackovic, T.M., Janev, M., Pilipovic, S., Zorica, D.: An expansion formula for fractional derivatives of variable order. Centr. Eur. J. Phys. **11**, 1350–1360 (2013)
5. Atanackovic, T.M., Janev, M., Konjik, S., Pilipovic, S., Zorica, D.: Expansion formula for fractional derivatives in variational problems. J. Math. Anal. Appl. **409**, 911–924 (2014)
6. Atanackovic, T.M., Stankovic, B.: An expansion formula for fractional derivatives and its application. Fract. Calc. Appl. Anal. **7**, 365–378 (2004)
7. Atanackovic, T.M., Stankovic, B.: On a numerical scheme for solving differential equations of fractional order. Mech. Res. Comm. **35**, 429–438 (2008)
8. Chen, Y., Ke, X., Wei, Y.: Numerical algorithm to solve system of nonlinear fractional differential equations based on wavelets method and the error analysis. Appl. Math. Comput. **251**, 475–488 (2015)
9. Diethelm, K., Ford, N.J., Freed, A.D.: A predictor-corrector approach for the numerical solution of fractional differential equations. Nonlinear Dynam. **29**, 3–22 (2002)
10. El-Sayed, A., Hashem, H., Ziada, E.: Picard and Adomian decomposition methods for a quadratic integral equation of fractional order. Comput. Appl. Math. **33**, 95–109 (2014)
11. Garrappa, R.: On linear stability of predictor-corrector algorithms for fractional differential equations. Int. J. Comput. Math. **87**, 2281–2290 (2010)
12. Li, C.P., Cai, M.: High-order approximation to Caputo derivatives and Caputo-type advection-diffusion equations: Revisited. Numer. Func. Anal. Optim. **38**, 861–890 (2017)
13. Li, C.P., Cai, M.: Theory and Numerical Approximations of Fractional Integrals and Derivatives. SIAM, Philadelphia (2019)
14. Li, C.P., Zeng, F.H.: Numerical Methods for Fractional Calculus. CRC Press, Boca Raton (2015)
15. Lynch, V.E., Carreras, B.A., Castillo-Negrete, D., Ferreira-Mejias, K.M., Hicks, H.R.: Numerical methods for the solution of partial differential equations of fractional order. J. Comput. Phys. **192**, 406–421 (2003)
16. Oldham, K.B., Spanier, J.: The Fractional Calculus. Academic Press, New York (1974)
17. Pooseh, S., Almeida, R., Torres, D.F.M.: Approximation of fractional integrals by means of derivatives. Comput. Math. Appl. **64**, 3090–3100 (2012)
18. Pooseh, S., Almeida, R., Torres, D.F.M.: Numerical approximations of fractional derivatives with applications. Asian J. Control **15**, 698–712 (2013)
19. Sousa, J.V., Machado, J.A.T., Oliveira, E.C.: The ψ-Hilfer fractional calculus of variable order and its applications. Comp. Appl. Math. **39**, 296 (2020)
20. Tavares, D., Almeida, R., Torres, D.F.M.: Caputo derivatives of fractional variable order: numerical approximations. Commun. Nonlinear Sci. Numer. Simul. **35**, 69–87 (2016)
21. Ziada, E.: Numerical solution for nonlinear quadratic integral equations. J. Fractional Calc. Appl. **7**, 1–11 (2013)

A Risk-Cost Analysis for an Increasingly Widespread Monitoring of Railway Lines

Imma Lory Aprea[1]([envelope]) [iD], Chiara Donnini[2] [iD], and Federica Gioia[2] [iD]

[1] Department of Economics, Law, Cybersecurity and Sports Sciences,
University of Naples "Parthenope", Guglielmo Pepe Street, 80035 Nola, Italy
immalory.aprea@uniparthenope.it
[2] Department of Business and Quantitative Studies, University of Naples
"Parthenope", Generale Parisi Street, 80132 Naples, Italy
{chiara.donnini,federica.gioia}@uniparthenope.it

Abstract. Structural Health Monitoring represents an essential tool for detecting timely failures that may cause potential damage to the railway infrastructure, such as extreme weather conditions, natural accidental phenomena, and heavy loads affecting tracks, bridges, and other structures over time. The more thorough the monitoring, the more exact the information can be derived. In this paper, we propose an optimal approach to ensure the maximum railway infrastructure reliability through increasingly widespread and effective monitoring, subject to a budget constraint. More in detail, considered a pre-existing network of zones, each of which is monitored by a set of fixed diagnostic sensors, our goal is to identify new additional areas in which to place the same set of sensors in order to evaluate the geometric and structural quality of the track simultaneously. A kriging technique is used to identify the riskiness of some unsampled locations in order to select the new areas to be monitored. Moreover, two different decision criteria have been introduced, both depending on the risk level of the occurrence of extreme phenomena under investigation and involving the analysis of monitoring and non-monitoring costs. A descriptive analysis of the procedure, which may be used to identify the additional zones to be monitored, is provided in the paper by illustrating the resolution algorithm of the problem. The methodology has been implemented in environment R by using simulated data.

Keywords: Risk-Cost Analysis · Sensor Networks · Structural Health Monitoring

1 Introduction

Structural Health Monitoring (SHM) represents an essential tool to timely detect potential railway faults that may cause damage or to prevent the scheduled ordinary operation. SHM provides information about the structural integrity of the railway system by processing data acquired by sensors [13]. Through sensor

H. Moosaei et al. (Eds.): DIS 2023, LNCS 14321, pp. 31–53, 2024.
https://doi.org/10.1007/978-3-031-50320-7_3

data, the behaviour of the most relevant mechanical and geometrical parameters of the system (temperature, train load, train dynamics, subgrade movements, etc.) can be analysed [2]. Therefore, sensor data suggest the actions required to maintain or recover the integrity of the entire infrastructure.

As highlighted in [14], one of the main causes of the railway infrastructure deterioration is represented by climate change. Indeed, the effects of climate change lead to a slow deterioration of the railway infrastructure and may affect performance significantly when extreme events occur. For instance, extreme temperatures, earthquakes, heavy precipitation, flooding, rising sea level, storm surge, and flash floods can crucially impact railway operations, reliability, safety, and cost. High temperatures may cause the buckled rail (which arises when metal rails expand due to hot weather [15–17,20]), power lines failures [24], drought and fire risks [22]. Furthermore, extreme temperature changes may lead to (or contribute to) broken rails, resulting often in derailments [6,7,18,19,21,23]. Rising sea level and storm surge may cause disruption to railway track near coastal areas; earthquakes, heavy precipitation, and flooding may lead to embankment and cutting slope failure; flash floods may result in washing away ballast [13]. Therefore, it is essential for railway companies to monitor the occurrence of extreme phenomena and take appropriate measures to reduce the risks associated with them.

In the literature, different methods have been discussed to optimally plan SHM in dynamic systems. The most known approaches address the optimal sensor allocation in a given system section, either maximising information quality or minimising costs. More specifically, in the first case, the number and type of sensors are assumed to be constant and the optimisation problem focuses on the spatial configuration and the choice of the objective function measuring the system reliability [3,4,9]. In the second case, a certain level of information quality is given and the optimization problem focuses on minimising the number of sensors, reducing the costs associated with the purchase, installation, and maintenance of sensors [1,8].

A different approach to optimally plan SHM addresses maximising system reliability through an increasingly widespread and effective monitoring. Such a monitoring represents a more challenging and still open issue for railway companies. In this case, the problem examines identifying additional areas along the railway line in which to allocate the sensors.

This paper is embedded in this branch of literature and provides a static risk-cost analysis for detecting new areas to be monitored. In particular, two different crucial problems, with respect to train safety, are analysed: the *thermal buckling* and the *cracks due to fatigue* (or *breaking down*) of rail. In detail, we search for a thicker network of controlled zones able to ensure the coverage of expenses and the maximum infrastructure reliability in terms of safety, social welfare, and limitation of environmental impacts.

Considering a pre-existing network of zones along a railway track section, each of which is monitored by a set of fixed diagnostic sensors, we aim to identify additional areas in which to place the same set of sensors to simultaneously evaluate the geometric and structural quality of the railway track. The detection of new zones is carried on through a risk-cost analysis. In particular, each new area is identified taking into account the monitoring and non-monitoring costs associated with it (where the non-monitoring cost represents a damage cost associated to the zone if at least one extreme phenomenon occurs). These two costs both are related to the estimated risk level associated with the extreme phenomena occurring in each zone. Moreover, the selection of the areas to be added to the pre-existing network is subject to a budget constraint. In this regard, we need to consider the physical parameters (related to the system to be monitored), the different types of sensors available in the market and their accuracy, the total number of sensors in each monitored zone, and the initial configuration. The probability distributions of the structural parameters to be estimated at unsampled points are inferred through the Universal Kriging interpolation technique [12]. This technique allows us to build a metamodel in order to assess the reliability of the railway system in the not monitored areas of the railway infrastructure [10].

The paper is structured as follows. In Sect. 2, the theoretical framework is presented. More specifically, we introduce the risk functions measuring the risk value for occurrence of each monitored phenomenon. In addition, we define the monitoring cost of each zone and the non-monitoring cost related to the damage cost to be supported if an extreme phenomenon occurs. In Sect. 3, the resolution algorithm of the whole process is provided, and two decision criteria, for selecting the additional zones to be monitored, are illustrated. In Sect. 4, a case study is presented, and the results obtained by the two decision criteria are compared. Finally, Sect. 5 contains the conclusion, while in Appendix the Kriging interpolation method is described.

2   Theoretical Framework

Let $\mathcal{A}$ be the area of a railway section bounded by two zones, $\mathcal{Z}_1$ and $\mathcal{Z}_{\bar{m}}$, equipped with a set of fixed diagnostic sensors in a system already in operation. Assume that $\mathcal{A}$ may be partitioned in $\bar{m}$ zones, m of which are monitored. Let us denote by $\mathcal{N}_{tot} = \{\mathcal{Z}_1, ..., \mathcal{Z}_{\bar{m}}\}$ the network of zones covering $\mathcal{A}$ and by $\mathcal{N} \subseteq \mathcal{N}_{tot}$ the subnet of pre-existing controlled zones, i.e., $\mathcal{N} = \{\mathcal{Z} \in \mathcal{N}_{tot} : \mathcal{Z} \text{ is monitored}\}$.

The monitoring system in the network $\mathcal{N}$ allows to check the possible occurrence of certain extreme phenomena that may affect the railway infrastructure, causing disruptions. Let $\mathbf{L} = \{L_1, \ldots, L_{n_\mathbf{L}}\}$ be the set of such possible extreme events. Let us denote by $\Theta = \{\theta_1, \ldots, \theta_{n_\Theta}\}$ the set collecting all parameters analysed by the system predisposed in $\mathcal{N}$. More precisely, Θ is composed of all structural parameters monitored in order to determine the risk level of occurrence for each phenomenon in $\mathbf{L}$. Each parameter in Θ contributes to the monitoring of at least one phenomenon to be controlled and each phenomenon in $\mathbf{L}$ is examined by analysing at least one parameter in Θ.

For each $\ell = 1, \ldots, n_\mathbf{L}$, we define the subset of Θ, whose elements are parameters contributing to monitor the phenomenon L_ℓ,

$$\boldsymbol{\Theta}_{L_\ell} = \{\theta_j : \theta_j \text{ contributes to monitor } L_\ell\},$$

while for each $j = 1, \ldots, n_\Theta$, we define the subset of $\mathbf{L}$ composed of the phenomena L_ℓ for monitoring which the parameter θ_j is analysed,

$$\mathbf{L}_{\theta_j} = \{L_\ell : L_\ell \text{ is analyzed by } \theta_j\}.$$

Under the previous considerations, we have that

- $\forall \ell = 1, \ldots, n_{\mathbf{L}}$, $|\boldsymbol{\Theta}_{L_\ell}| \geq 1$, that is the subset $\boldsymbol{\Theta}_{L_\ell}$ has at least one type of parameter;
- $\bigcup_{\ell=1}^{n_{\mathbf{L}}} \boldsymbol{\Theta}_{L_\ell} = \boldsymbol{\Theta}$, namely, the union of all subsets $\boldsymbol{\Theta}_{L_\ell}$ returns the entire set $\boldsymbol{\Theta}$;
- $\{\boldsymbol{\Theta}_{L_1}, \ldots, \boldsymbol{\Theta}_{L_{n_{\mathbf{L}}}}\}$ is not necessarily a partition of $\boldsymbol{\Theta}$;
- $\forall j = 1, \ldots, n_\Theta$, $|\mathbf{L}_{\theta_j}| \geq 1$, specifically the subset $\mathbf{L}_{\theta_j}$ has at least one type of risk L_ℓ.

According to the protocol established for monitoring zones in the area $\mathcal{A}$, there is a one to one correspondence between parameters and typology of sensors that allows to detect their values. In addition, in each controlled (or to be controlled) area, the number of sensors used (or to be used) to monitor each parameter is assumed to be fixed. In other words, for each $j = 1, \ldots, n_\Theta$, there exists a unique type of sensor s_j monitoring the parameter θ_j and each type of sensor s_j monitors a unique parameter θ_j. Furthermore, in each controlled (or to be controlled) zone, for each θ_j, n_{θ_j} sensors s_j (with $n_{\theta_j} \geq 1$) have to be allocated to detect the value that θ_j assumes in the zone. Let us denote by $n = \sum_{j=1}^{n_\Theta} n_{\theta_j}$ the total number of sensors used in each monitored zone.

Risk Functions Associated with Possible Accidental Phenomena
As previously observed, fixed a phenomenon L_ℓ, each structural parameter in $\boldsymbol{\Theta}_{L_\ell}$ contributes to evaluate the possible occurrence of L_ℓ. Therefore, starting from the relationship between the parameter and the phenomenon, it is possible to construct a risk measure evaluating the occurrence of L_ℓ depending on the probability distribution of each element of $\boldsymbol{\Theta}_{L_\ell}$.

Let us denote by $\mathbb{P}$ the space of probability distributions.

For each $\ell = 1, \ldots, n_{\mathbf{L}}$, for each parameter $\theta_j \in \boldsymbol{\Theta}_{L_\ell}$, we denote by

$$\mathcal{R}_{j,\ell} : \mathbb{P} \to \mathbb{N} \cup 0,$$

the risk measure associating to each probability distribution function $p(\theta)$ the value $\mathcal{R}_{j,\ell}(p(\theta))$, i.e., the risk level for occurrence of L_ℓ if the probability distribution function of the parameter θ_j is assumed to be equal to $p(\theta)$.

Cost of Monitoring a Zone
Monitoring a zone allows to control the possible occurrence of a phenomenon, ensuring timely, rapid, and prompt intervention if needed. Knowing in advance that a phenomenon could happen, in fact, enables one to choose to implement

strategies that lead to counteract the event and/or its effects. On the other hand, monitoring involves an expense. In addition to fixed costs, which include, for instance, data acquisition hardware, databases, and assemblages, different types of variable costs have to be computed for each zone. As previously observed, for each parameter, the protocol provides a single type of sensor. However, in the market there are sensors of the same type, but of different quality. In choosing the quality of the sensor to be installed, the accuracy deemed necessary for the analysis plays a fundamental role. The greater the estimated risk level of occurrence of a phenomenon, the higher the level of accuracy required by the sensor detecting data of a parameter contributing to monitor the phenomenon. As a consequence, the greater the estimated risk level of occurrence of a phenomenon, the higher the cost of the sensor. Namely, we are assuming that, in the sensor selection criterion, there exist a one to one increasing correspondence between risk level and accuracy and a one to one increasing correspondence between accuracy and cost. Hence, we can consider, for each $j = 1, ..., n_\Theta$, the cost of the sensor s_j to be used, C_j, directly depending on the estimated risk level of occurrence of a phenomenon L_ℓ belonging to $\mathbf{L}_{\theta_j}$. Formally, $C_j\left(\mathcal{R}_{j,\ell}(p(\theta_j))\right)$ represents the unitary cost per sensor s_j depending on the accuracy needed to monitor the parameter θ_j, given the risk value $\mathcal{R}_{j,\ell}(p(\theta_j))$.

As previously observed, a parameter θ_j may be analysed to estimate the risk level of occurrence for several phenomena. Therefore, in selecting the quality of sensors s_j to be placed, it is necessary to focus on the highest risk level among those associated with the phenomena that θ_j contributes to monitoring.

Given a zone $\mathcal{Z}$, for each $j = 1, ..., n_\Theta$, let us denote by $p(\theta_{j,\mathcal{Z}})$ the probability distribution of θ_j estimated in $\mathcal{Z}$. Let us define

$$\mathcal{R}_j^*(\mathcal{Z}) = \max_{L_\ell \in \mathbf{L}_{\theta_j}} \left\{\mathcal{R}_{j,\ell}(p(\theta_{j,\mathcal{Z}}))\right\}.$$

Therefore, the cost of monitoring $\mathcal{Z}$ may be decomposed in two addenda, the first representing the fixed cost, C, and the second involving the sum of the costs of all sensors to be installed, that is,

$$\mathcal{C}(\mathcal{Z}) = C + \sum_{j=1}^{n_\Theta} n_{\theta_j} C_j\left(\mathcal{R}_j^*(\mathcal{Z})\right).$$

Cost of Non-monitoring a Zone
Deciding not to monitor a zone could lead to damage costs, if even at least one of the phenomena that should have been controlled occurred. Indeed, the occurrence of an adverse phenomenon could lead to an intervention cost due to the damage, a cost given by the interruption of the service, and other related costs. For each zone $\mathcal{Z}$ and for each $\ell = 1, ..., n_\mathbf{L}$, it is, therefore, possible to evaluate the cost of damage to face whether the phenomenon L_ℓ occurs or not in the zone $\mathcal{Z}$. Let us denote by $\mathcal{D}(L_{\ell,\mathcal{Z}})$ such value and observe that it can be decomposed in the sum of costs independent from the zone (for instance, spare parts costs) and costs depending on the zone (for example, the more the area is

inaccessible, the longer the time will be needed for repair, the higher will be the costs due to the interruption of service).

For each zone $\mathcal{Z}$ and each phenomenon L_ℓ, denoting by $\mathcal{P}_{j,\ell}(p(\theta_{j,\mathcal{Z}}))$ the probability that L_ℓ occurs in $\mathcal{Z}$, we define

$$\mathcal{P}_\ell^*(\mathcal{Z}) = \max_{\theta_j \in \Theta_{L_\ell}} \{\mathcal{P}_{j,\ell}(p(\theta_{j,\mathcal{Z}}))\}$$

as the highest probability that the phenomenon L_ℓ is realised, among those estimated by the analysis of each parameter θ_j in Θ_{L_ℓ}. Therefore, for any zone $\mathcal{Z}$, we define the cost of non-monitoring it, $\mathcal{D}(\mathcal{Z})$, as

$$\mathcal{D}(\mathcal{Z}) = \sum_{\ell=1}^{n_L} \mathcal{P}_\ell^*(\mathcal{Z})\mathcal{D}(L_{\ell,\mathcal{Z}}).$$

Budget Constraint

The aim of our investigation is to obtain a thicker network of monitored zones in order to increase the infrastructure reliability in terms of safety and social welfare. The greater the number of controlled zones and the better the considered accuracy, the higher the efficiency of the service provided by the railway companies. However, as previously observed, monitoring new zones leads costs. The choice of the additional areas must take into account a budget constraint.

If B is the available budget, the new network $\mathcal{N}^* \subseteq \mathcal{N}_{tot} \setminus \mathcal{N}$ has to satisfy the following budget constraint,

$$B \geq \sum_{\mathcal{Z}^* \in \mathcal{N}^*} \mathcal{C}(\mathcal{Z}^*).$$

3   Risk-Cost Analysis

The problem of identifying additional areas along the railway track requires an analysis of monitoring and non-monitoring costs. This analysis can not disregard the estimation of the probability distribution of parameters in points where the area is not monitored. Indeed, in order to compute the amount of costs in both cases, it is first necessary to fill the lack of data information due to the absence of sensors in some points of the selected area of the railway, $\mathcal{A}$.

A first part of our analysis, therefore, concerns the estimation of the missing data and the corresponding distribution probability functions. Once these estimates have been obtained, it will therefore be possible to assess, at every point of the area under consideration, the level of risk for occurrence of each phenomenon to be monitored. As a consequence, for each possible zone, the costs of monitoring and non-monitoring the zone can be computed and compared in order to make the optimal choice in terms of cost and safety.

In the following, we briefly describe step by step the analysis of the optimal identification of new areas to be monitored.

Estimation of Values and Distribution Functions
For any $i = 1, ..., m$, $j = 1, ..., n_\Theta$ and $k = 1, ..., n_{\theta_j}$, let us denote by $P^i_{k,j} \in \mathcal{A}$ the point of the zone $\mathcal{Z}_i$ where the k-th sensor s_j is positioned.

For any $j = 1, ..., n_\Theta$, we collect data of the parameter θ_j observed in the points $P^i_{k,j}$ with $i = 1, ..., m$ and $k = 1, ..., n_{\theta_j}$.

Using the Kriging interpolation method[1], we estimate the values of θ_j in each non-monitored point of $\mathcal{A}$. More precisely, considering as input data

- the observed and previously collected data,
- the coordinates of the points where they were observed,
- the area under analysis,
- the appropriate calculation grid,

we obtain as output the estimated values of the parameter θ_j in each element of the grid.

The calculation grid given as input in the Kriging interpolation method allows to consider, with good approximation, each zone $\mathcal{Z}$ in $\mathcal{N}_{tot}$ partitioned into a natural number of grid cells, hence, starting from the obtained estimations, with a standard method, we estimate the distribution function of θ_j in each zone $\mathcal{Z} \in \mathcal{N}_{tot}$.

Evaluation of the Risk Level for Occurrence of the Phenomena
For each L_ℓ in $\mathbf{L}_{\theta_j}$, starting from the distribution functions obtained at the previous step, we estimate the risk level for occurrence of L_ℓ.

Considering as input data

- the functions relating each parameter θ_j to L_ℓ and the conditions establishing the threshold beyond which the extreme phenomenon is realised,
- the distribution functions estimated at the previous step,
- the risk functions $\mathcal{R}_{j,\ell}$,

we obtain, for each zone $\mathcal{Z} \in \mathcal{N}_{tot}$, the risk levels for the occurrence of L_ℓ, estimated through the probability distribution of the parameter θ_j, i.e., $\mathcal{R}_{j,\ell}(p(\theta_{j,\mathcal{Z}}))$.

Through standard methods, we produce the chromatic maps of the area representing the risk levels of occurrence of each phenomenon L_ℓ in $\mathbf{L}_{\theta_j}$.

Since the procedure is implemented for each parameter θ_j, we get $\sum_{j=1}^{n_\Theta} |\mathbf{L}_{\theta_j}|$ maps. Indeed, we obtain the chromatic maps related to every phenomenon through each of the parameters that help to monitor it.

In other word, we identify the estimated risk level for occurrence of each phenomenon related to each parameter contributing to monitoring it. As a consequence, for each possible zone, cost of monitoring and cost of non-monitoring may be computed and considered for decision-making. In order to be selected, a new network has to satisfy the budget constraint.

---

[1] See Appendix for a brief description of the Kriging interpolation method.

Selection Criteria

In the following we propose two different selection criteria, both taking into account monitoring and non-monitoring costs and accuracy of monitoring.

Before describing them and providing their algorithms, we need to introduce some notations.

For each zone $\mathcal{Z}$ in $\mathcal{N}_{tot}$, for each $j = 1, ..., n_\Theta$ and $k \in \mathbb{N} \cup \{0\}$, we define

$$\mathcal{C}^k(\mathcal{Z}) = C + \sum_{j=1}^{n_\Theta} n_{\theta_j} C_j \left((\mathcal{R}_j^*(\mathcal{Z}) - k)^+ \right),$$

where $(\mathcal{R}_j^*(\mathcal{Z}) - k)^+ = \max\{\mathcal{R}_j^*(\mathcal{Z}) - k, 0\}$. Notice that $\mathcal{C}^0(\mathcal{Z}) = \mathcal{C}(\mathcal{Z})$, while for every $k \in \mathbb{N}$, $\mathcal{C}^k(\mathcal{Z})$ identifies the cost that we have to face if we decide to give up some of accuracy, using sensors less accurate than necessary. Indeed, choosing to place sensors s_j which unitary cost is $C_j \left((\mathcal{R}_j^*(\mathcal{Z}) - k)^+ \right)$, we choose to lower $\min \mathcal{R}_j^*(\mathcal{Z})$, k degree of accuracy.

With abuse of notation, given a network $\tilde{\mathcal{N}} \subseteq \mathcal{N}_{tot}$, for each $k \in \mathbb{N} \cup \{0\}$, we denote

$$\mathcal{C}^k(\tilde{\mathcal{N}}) = \sum_{\mathcal{Z} \in \tilde{\mathcal{N}}} \mathcal{C}^k(\mathcal{Z}).$$

Given the network $\mathcal{N}^0 = \mathcal{N}_{tot} \setminus \mathcal{N}$, we define $\bar{\alpha} = \max_{\mathcal{Z} \in \mathcal{N}^0} \{\mathcal{R}_j^*(\mathcal{Z}), j = 1, ..., n_\Theta\}$ and for each $\alpha = 1, ..., \bar{\alpha}$,

- $\bar{\mathcal{N}}^\alpha = \left\{ \mathcal{Z} \in \mathcal{N}^0 : \max_{\theta_j \in \Theta} \mathcal{R}_j^*(\mathcal{Z}) \geq \alpha \right\}$, i.e., the network composed of all zones with maximal risk at least α;
- $\bar{\mathcal{N}}_-^\alpha = \{ \mathcal{Z} \in \bar{\mathcal{N}}^\alpha : \mathcal{C}(\mathcal{Z}) > \mathcal{D}(\mathcal{Z}) \}$, i.e., the subnet of $\bar{\mathcal{N}}^\alpha$ collecting all zones with non-monitoring cost less than monitoring one;
- $\mathcal{N}^\alpha = \left\{ \mathcal{Z} \in \mathcal{N}^0 : \max_{\theta_j \in \Theta} \mathcal{R}_j^*(\mathcal{Z}) = \alpha \right\}$, i.e., the network composed of all zones with maximal risk equal to α;
- $\mathcal{N}_-^\alpha = \{ \mathcal{Z} \in \mathcal{N}^\alpha : \mathcal{C}(\mathcal{Z}) > \mathcal{D}(\mathcal{Z}) \}$, i.e., all zones with maximal risk level equal to α and non-monitoring cost less than monitoring cost;
- for each $h = \alpha, ..., \bar{\alpha}$,
  - $\hat{\mathcal{N}}_-^h = \mathcal{N}_-^h \cup \left(\bigcup_{\alpha-1 \leq k < h} \mathcal{N}^k \right)$, i.e., the network collecting all zones with maximal risk level equal to h and non-monitoring cost less than monitoring one, and all zones with maximal risk level greater than or equal to $\alpha - 1$ and less than h;
  - $\hat{\mathcal{N}}^h = \bigcup_{\alpha-1 \leq k \leq h} \mathcal{N}^k$, i.e., the network composed of all zones with maximal risk level greater than or equal to $\alpha - 1$ and less than or equal to h.

First Criterion

The first criterion is composed by two parts. In the former, the goal is to find, if there exists, the minimum α belonging to $\{1, ..., \bar{\alpha}\}$ for which the budget is sufficient to monitor $\bar{\mathcal{N}}^{\alpha}$. In the latter, which has to be implemented only if the first part fails, it is studied if it is possible to control at least zones with monitoring cost less than the non-monitoring one and maximal risk level $\bar{\alpha}$.

Let us describe the criterion more in detail.

For each $\alpha \in \{1, ..., \bar{\alpha}\}$, consider the network $\bar{\mathcal{N}}^{\alpha}$ in which each zone is equipped, for each parameter θ_j, by sensors s_j with cost $C_j((\mathcal{R}_j^*(\mathcal{Z}) - \alpha + 1)^+)$, i.e., with accuracy lowered by $\alpha - 1$ degrees.

If the budget is sufficient, try to improve the accuracy of sensors as much as possible. Namely, if $\mathcal{C}^{\alpha-1}(\bar{\mathcal{N}}^{\alpha}) \leq B$, determine $\beta = \max\{\gamma \in \{1, ..., \alpha\} : \mathcal{C}^{\alpha-\gamma}(\bar{\mathcal{N}}^{\alpha}) \leq B\}$ and fix $\mathcal{N}^* = \bar{\mathcal{N}}^{\alpha}$, considering in each $\mathcal{Z} \in \bar{\mathcal{N}}^{\alpha}$, for every parameter θ_j, sensors s_j with unitary cost $C_j((\mathcal{R}_j^*(\mathcal{Z}) - \alpha + \beta)^+)$. The process is stopped at the first α satisfying the requirement.

Otherwise, if the budget is not enough starting from $h = \alpha$, lower the accuracy level of the sensors positioned in all zones of $\bar{\mathcal{N}}^{\alpha}$ with maximum risk $h - 1$ and in those ones with maximum risk h and non monitoring cost lower than the monitoring one. If the budget is sufficient, consider $\mathcal{N}^* = \bar{\mathcal{N}}^{\alpha}$, in which, for each parameter θ_j, each zone $\mathcal{Z}$ in $\bar{\mathcal{N}}^{\alpha} \setminus \hat{\mathcal{N}}_-^h$ is monitored by sensors s_j with unitary cost $C_j((\mathcal{R}_j^*(\mathcal{Z}) - \alpha + 1)^+)$, while each zone $\mathcal{Z}$ in $\hat{\mathcal{N}}_-^h$ is monitored by sensors s_j with unitary cost $C_j((\mathcal{R}_j^*(\mathcal{Z}) - \alpha)^+)$. If the previous operation is still not enough, lower the accuracy level even to all zones with maximum risk h and non monitoring cost greater than or equal to the monitoring cost. If it is sufficient, then consider $\mathcal{N}^* = \bar{\mathcal{N}}^{\alpha}$, in which, for each parameter θ_j, each zone $\mathcal{Z}$ in $\bar{\mathcal{N}}^{\alpha} \setminus \hat{\mathcal{N}}^h$ is monitored by sensors s_j with unitary cost $C_j((\mathcal{R}_j^*(\mathcal{Z}) - \alpha + 1)^+)$, while each zone $\mathcal{Z}$ in $\hat{\mathcal{N}}^h$ is monitored by sensors s_j with unitary cost $C_j((\mathcal{R}_j^*(\mathcal{Z}) - \alpha)^+)$.

If, using sensors with cost $C_j(\mathcal{R}_j^*(\mathcal{Z}) - \bar{\alpha})$, the budget is not enough for monitoring $\bar{\mathcal{N}}^{\bar{\alpha}}$, check if the budget constraint is satisfied considering only zones with maximal risk $\bar{\alpha}$ and non-monitoring cost less than monitoring cost. If it is not possible, declare the budget insufficient.

The algorithm for the criterion is presented in the following.

- For each $\alpha = 1, ..., \bar{\alpha}$, compute $\mathcal{C}^{\alpha-1}(\bar{\mathcal{N}}^{\alpha})$.
  - if $\mathcal{C}^{\alpha-1}(\bar{\mathcal{N}}^{\alpha}) \leq B$,
    compute $\beta = \max\{\gamma \in \{1, ..., \alpha\} : \mathcal{C}^{\alpha-\gamma}(\bar{\mathcal{N}}^{\alpha}) \leq B\}$ and consider $\mathcal{N}^* = \bar{\mathcal{N}}^{\alpha}$ in which, for each parameter θ_j, each zone $\mathcal{Z}$ is monitored by sensors s_j with unitary cost $C_j((\mathcal{R}_j^*(\mathcal{Z}) - \alpha + \beta)^+)$.
  - if $\mathcal{C}^{\alpha-1}(\bar{\mathcal{N}}^{\alpha}) > B$,
    for each $h = \alpha, ..., \bar{\alpha}$,
    * if $\mathcal{C}^{\alpha-1}(\bar{\mathcal{N}}^{\alpha} \setminus \hat{\mathcal{N}}_-^h) \leq B$, consider $\mathcal{N}^* = \bar{\mathcal{N}}^{\alpha}$ in which, for each parameter θ_j, each zone $\mathcal{Z}$ in $\bar{\mathcal{N}}^{\alpha} \setminus \hat{\mathcal{N}}_-^h$ is monitored by sensors s_j with unitary cost $C_j((\mathcal{R}_j^*(\mathcal{Z}) - \alpha + 1)^+)$, while each zone $\mathcal{Z}$ in $\hat{\mathcal{N}}_-^h$ is monitored by sensors s_j with unitary cost $C_j((\mathcal{R}_j^*(\mathcal{Z}) - \alpha)^+)$.

∗ if $\mathcal{C}^{\alpha-1}(\bar{\boldsymbol{N}}^\alpha \setminus \hat{\boldsymbol{N}}_-^h) > B$, compute $\mathcal{C}^{\alpha-1}(\bar{\boldsymbol{N}}^\alpha \setminus \hat{\boldsymbol{N}}^h)$.

  · if $\mathcal{C}^{\alpha-1}(\bar{\boldsymbol{N}}^\alpha \setminus \hat{\boldsymbol{N}}^h) \leq B$,

    consider $\boldsymbol{N}^* = \bar{\boldsymbol{N}}^\alpha$ in which, for each parameter θ_j, each zone $\mathcal{Z}$ in $\bar{\boldsymbol{N}}^\alpha \setminus \hat{\boldsymbol{N}}^h$ is monitored by sensors s_j with unitary cost $C_j((\mathcal{R}_j^*(\mathcal{Z}) - \alpha + 1)^+)$, while each zone $\mathcal{Z}$ in $\hat{\boldsymbol{N}}^h$ is monitored by sensors s_j with unitary cost $C_j((\mathcal{R}_j^*(\mathcal{Z}) - \alpha)^+)$.

  · if $\mathcal{C}^{\alpha-1}(\bar{\boldsymbol{N}}^\alpha \setminus \hat{\boldsymbol{N}}^h) > B$, go on for $h + 1$.

∗ if $\mathcal{C}^{\alpha-1}(\bar{\boldsymbol{N}}^\alpha \setminus \hat{\boldsymbol{N}}_-^\alpha) > B$, go on for $\alpha + 1$.

– if $\mathcal{C}^{\bar{\alpha}}(\bar{\boldsymbol{N}}^{\bar{\alpha}}) > B$,
  compute $\mathcal{C}^{\bar{\alpha}}(\bar{\boldsymbol{N}}^{\bar{\alpha}} \setminus \boldsymbol{N}_-^{\bar{\alpha}})$

  • if $\mathcal{C}^{\bar{\alpha}}(\bar{\boldsymbol{N}}^{\bar{\alpha}} \setminus \boldsymbol{N}_-^{\bar{\alpha}}) \leq B$, consider $\boldsymbol{N}^* = \bar{\boldsymbol{N}}^{\bar{\alpha}} \setminus \boldsymbol{N}_-^{\bar{\alpha}}$ in which, for each parameter θ_j, each zone $\mathcal{Z}$ is monitored by sensors s_j with unitary cost $C_j((\mathcal{R}_j^*(\mathcal{Z}) - \bar{\alpha})^+)$,

  • if $\mathcal{C}^{\bar{\alpha}}(\bar{\boldsymbol{N}}^{\bar{\alpha}} \setminus \boldsymbol{N}_-^{\bar{\alpha}}) > B$, declare budget insufficient.

Second Criterion

For each $\alpha \in \{1, ..., \bar{\alpha}\}$ consider the network $\boldsymbol{N}^\alpha$ in which, for each parameter θ_j, each zone is equipped by sensors s_j with unitary cost $C_j(\mathcal{R}_j^*(\mathcal{Z}))$. Check if there exists the minimum α for which the budget is sufficient to monitoring $\boldsymbol{N}^\alpha$, with maximum required accuracy.

If the budget is not enough, check if there exists the minimum α for which the budget is sufficient to monitoring all zones in $\boldsymbol{N}^\alpha$ except those for which the non-monitoring cost is less than the monitoring one.

If the budget is not sufficient even after such a reduction in the number of zones, lower the accuracy level of the sensors as long as the budget is enough, and then try, if it is possible, to add zones considering those with non-monitoring cost less than monitoring one. If it is not possible to control all zones with maximal risk $\bar{\alpha}$ using, for each parameter θ_j, sensors s_j with minimum cost, declare the budget insufficient.

Otherwise, if the budget is sufficient to monitoring all zones in $\boldsymbol{N}^\alpha \setminus \boldsymbol{N}_-^\alpha$ with the maximum required accuracy, try to increase the number of zones to be monitored, adding, for any $h = 0, ..., \alpha$, all zones with maximal risk less than or equal to $\alpha - h$ and non-monitoring cost less than monitoring one.

The algorithm for the criterion is presented in the following.

For any $\alpha = 1, ..., \bar{\alpha}$, compute $\mathcal{C}(\boldsymbol{N}^\alpha)$.

– if $\mathcal{C}(\boldsymbol{N}^\alpha) \leq B$,
  consider $\boldsymbol{N}^* = \boldsymbol{N}^\alpha$ in which each zone is monitored with sensors that guarantee the maximum accuracy required.

– if $\mathcal{C}(\boldsymbol{N}^\alpha) > B$,
  compute $\mathcal{C}(\boldsymbol{N}^\alpha \setminus \boldsymbol{N}_-^\alpha)$.

– if $\mathcal{C}(\boldsymbol{N}^\alpha \setminus \boldsymbol{N}_-^\alpha) > B$, go on for $\alpha + 1$.

- if $\mathcal{C}(\bar{\mathcal{N}}^{\bar{\alpha}} \setminus \bar{\mathcal{N}}_-^{\bar{\alpha}}) > B$,

  for any $k = 1, ..., \bar{\alpha}$, compute $\mathcal{C}^k(\bar{\mathcal{N}}^{\bar{\alpha}} \setminus \bar{\mathcal{N}}_-^{\bar{\alpha}})$.

  * if $\mathcal{C}^k(\bar{\mathcal{N}}^{\bar{\alpha}} \setminus \bar{\mathcal{N}}_-^{\bar{\alpha}}) > B$, go on for $k + 1$.

    · if $\mathcal{C}^{\bar{\alpha}}(\bar{\mathcal{N}}^{\bar{\alpha}} \setminus \bar{\mathcal{N}}_-^{\bar{\alpha}}) > B$, declare the budget insufficient.

  * if $\mathcal{C}^k(\bar{\mathcal{N}}^{\bar{\alpha}} \setminus \bar{\mathcal{N}}_-^{\bar{\alpha}}) \le B$,

    for any $h = 1, ..., \bar{\alpha} - 1$, compute $\mathcal{C}^k(\bar{\mathcal{N}}^{\bar{\alpha}-h} \setminus \bar{\mathcal{N}}_-^{\bar{\alpha}-h})$.

    · if $\mathcal{C}^k(\bar{\mathcal{N}}^{\bar{\alpha}-h} \setminus \bar{\mathcal{N}}_-^{\bar{\alpha}-h}) \le B$, go on for $k + 1$.

    · if $\mathcal{C}^k(\bar{\mathcal{N}}^{\bar{\alpha}-h} \setminus \bar{\mathcal{N}}_-^{\bar{\alpha}-h}) > B$,

      consider $\mathcal{N}^* = \bar{\mathcal{N}}^{\bar{\alpha}-h+1} \setminus \bar{\mathcal{N}}_-^{\bar{\alpha}-h+1}$ in which, for each θ_j, each zone is equipped with sensors s_j which cost is $C_j((\mathcal{R}_j^*(\mathcal{Z}) - k)^+)$.

- if $\mathcal{C}(\bar{\mathcal{N}}^\alpha \setminus \bar{\mathcal{N}}_-^\alpha) \le B$,

  for any $h = 0, ..., \alpha - 1$, compute $\mathcal{C}(\bar{\mathcal{N}}^\alpha \setminus \bar{\mathcal{N}}_-^\alpha) + \mathcal{C}\left(\bigcup_{0 \le k \le h} \bar{\mathcal{N}}_-^{\alpha-k} \right)$.

- if $\mathcal{C}(\bar{\mathcal{N}}^\alpha \setminus \bar{\mathcal{N}}_-^\alpha) + \mathcal{C}\left(\bigcup_{0 \le k \le h} \bar{\mathcal{N}}_-^{\alpha-k} \right) \le B$, go on for $h + 1$.

- if $\mathcal{C}(\bar{\mathcal{N}}^\alpha \setminus \bar{\mathcal{N}}_-^\alpha) + \mathcal{C}\left(\bigcup_{0 \le k \le h} \bar{\mathcal{N}}_-^{\alpha-k} \right) > B$,

  consider $\mathcal{N}^* = (\bar{\mathcal{N}}^\alpha \setminus \bar{\mathcal{N}}_-^\alpha) \bigcup_{0 \le k \le h-1} \bar{\mathcal{N}}_-^{\alpha-k}$ in which each zone is monitored with sensors that guarantee the maximum accuracy required.

4   Case Study

We validate our risk-cost analysis by using simulated data of the rail track temperature. More precisely, we consider a tangent-track railway section $\mathcal{A}$ of length $\ell_{\mathcal{A}} = 2\,400$ m and height[2] $h_{\mathcal{A}} = 3$ m, along which there are 30 sensors measuring the rail track temperature. The section $\mathcal{A}$ can be considered partitioned in $\bar{m} = 50$ zones $\mathcal{Z}_i$, $i = 1, ..., 50$, both monitored and not, each of which has length[3] $\ell_{\mathcal{Z}} = 48$ m and height $h_{\mathcal{Z}} = 3$ m.

The pre-existing network is given by

$$\mathcal{N} = \{ \mathcal{Z}_1, \mathcal{Z}_6, \mathcal{Z}_{12}, \mathcal{Z}_{17}, \mathcal{Z}_{21}, \mathcal{Z}_{27}, \mathcal{Z}_{32}, \mathcal{Z}_{39}, \mathcal{Z}_{44}, \mathcal{Z}_{50} \}.$$

---

[2] The height of the examined railway section is established according to the rail gauge, i.e., the distance between the two rails of the same track. For instance, in Italy, this distance utilised by Rete Ferroviaria Italiana (RFI), the Manager of the Italian Railway Infrastructure, is fixed at $1\,345$ m.

[3] The length of the single zone has been selected among some reference values, established by a convergence analysis that detected the minimum length of the track model below which the track response variation is lower than a prescribed value (see [15] for more details).

The distances, expressed in meters, among a controlled zone and the subsequent one are collected in the vector $\boldsymbol{d} \in \mathbb{R}^9$ as follows,

$$\boldsymbol{d} = (192, 240, 192, 144, 240, 192, 288, 192, 240).$$

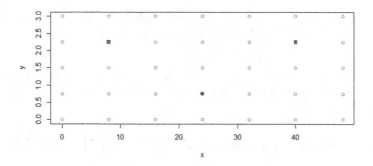

Fig. 1. Positions of sensors in the first monitored zone $\mathcal{Z}_1$.

As illustrated in Fig. 1, representing the calculation grid covering the zone $\mathcal{Z}_1$, each zone $\mathcal{Z}_i$ can be partitioned in 35 cells, whose 5 are shared with both the two adjacent zones. The total number of points into which the grid is partitioned is 1 505.

We test our approach by using $n_\theta = 1$ parameter θ, that is the rail track temperature, and $n_\ell = 2$ phenomena, for the monitoring of which the same parameter temperature is analysed: the *buckling phenomenon*, i.e., the zigzag pattern of the tracks due to high temperatures, and the *rail breaking down phenomenon* potentially caused by low temperatures. In each zone, we assume to have $n = 3$ thermometers, always in the same positions, as represented by the three red points in Fig. 1.

The proposed case study represents a simplified case aimed at testing the risk-cost analysis algorithms previously described. Therefore, in this framework, the examined area is assumed to be a tangent-track railway section with no curved-track[4].

We propose a static analysis of the rail track temperature monitoring by using the daily data of minimum, medium and maximum rail track temperature acquired by each of the 30 sensors for a year. The analysis has been carried out by using the software environment R for statistical computing.

---

[4] More generally, the railway track section to be analysed can be tangent, curved or partially tangent and partially curved. According to the case treated, the threshold values determining the occurrence of a given phenomenon change. Furthermore, in the case of a curved-track, these thresholds change according to the curve radius value, as well. For instance, in the case of the buckling phenomenon, the minimum sill temperature of buckling decreases as the curve radius increases. For more details, see [15].

4.1   Data Interpolation and Goodness-of-Fit

Firstly, for each data set (minimum, medium and maximum temperatures), we apply the Universal Kriging technique to estimate the values of the rail track temperature for each spatial point of the grid covering the area $\mathcal{A}$. For the variogram function, a penta spherical model has been used, suitable to interpolate spatial data on a daily basis [11].

Table 1. NSE and RMSE of the interpolated temperature data for the different temperature data sets.

Data set	NSE	RMSE
Minimum Temperatures	0.7569	4.1124
Medium Temperatures	0.7926	3.9412
Maximum Temperatures	0.7617	4.0934

The estimates provided by the interpolation process are reliable and sufficiently accurate, as confirmed by the obtained values of some statistical measures of goodness-of-fit (see Table 1). In detail, we compute the Nash Sutcliffe Efficiency (NSE) and the Root Mean Square Error (RMSE) for each data set (minimum, medium and maximum temperatures). It should be noted that the NSE returns values sufficiently high and almost close to 1 (NSE maximum value) in all three cases (0.7569, 0.7926 and 0.7617, respectively), proving that the estimated model is a good predictor. On the other hand, the RMSE maintains stable around low values (4.1124, 3.9412 and 4.0934, respectively). The most accurate estimates are achieved with the medium temperatures with an NSE $= 0.7926$ and an RMSE $= 3.9412$.

As a result, all the temperature data (minimum, medium and maximum ones) provided by the interpolation processes and related to a zone have been later employed to estimate the empirical probability distribution of the rail track temperature in the zone. By iterating this process for all the zones of the area $\mathcal{A}$, we obtain 50 probability distribution functions $p(\theta_{\mathcal{Z}_i})$, $i = 1, \ldots, 50$, each of which is used to determine the risk level $\mathcal{R}_\ell(p(\theta_{\mathcal{Z}_i}))$, $\ell \in \{1, 2\}$, for the occurrence of the buckling and breaking phenomena in the zone $\mathcal{Z}_i$.

4.2   Risk Level for the Buckling and Breaking Phenomena

The buckling phenomenon represents a circumstance in which the rail tracks change their shape, assuming a weaving side-to-side movement. This problem can be caused if the rail temperature increases above a certain critical value [16]. More precisely, there are two buckling critical values: a minimum temperature of buckling and a maximum temperature of buckling[5]. The minimum temperature of buckling, T_{buck}^{MIN}, represents the value below which the probability that

---

[5] See [15] for a detailed discussion.

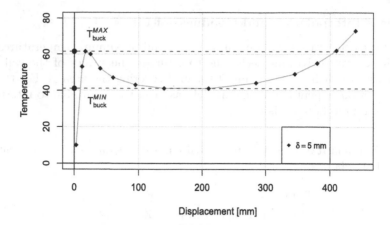

Fig. 2. Minimum and maximum buckling temperatures for a tangent-track with alignment defect $\delta = 5$ mm.

the phenomenon occurs is equal to 0. The maximum temperature of buckling, T_{buck}^{MAX}, is the value above which the rail track temperature is extremely high that the phenomenon happens. Figure 2 illustrates the minimum and maximum temperatures of buckling for a tangent-track with an alignment defect $\delta = 5$ mm, where $T_{\text{buck}}^{MIN} = 41\,°\text{C}$ and $T_{\text{buck}}^{MAX} = 61.5\,°\text{C}$.

Denoted by p_{buck} the probability that the buckling occurs according to the analysis of the parameter θ, this probability can be computed as

$$p_{\text{buck}} = p(\theta > T_{\text{buck}}^{MIN}).$$

Given the examined phenomenon, we assign

$$\mathcal{R}_{\text{buck}}(\mathcal{Z}) = \begin{cases} 0 & \text{if} & p_{\text{buck}} = 0 \\ 1 & \text{if} & 0 < p_{\text{buck}} \le 0.20 \\ 2 & \text{if} & 0.20 < p_{\text{buck}} \le 0.35 \\ 3 & \text{if} & 0.35 < p_{\text{buck}} \le 0.54 \\ 4 & \text{if} & p_{\text{buck}} > 0.54. \end{cases}$$

Figure 3 shows the colour thermal map of the risk level related to the buckling phenomenon for each zone of the area $\mathcal{A}$, obtained by applying the criterion previously described. In the figure, the blue vertical lines delimit the controlled zones of the pre-existing network. It should be noted that there are no zones with risk levels equal to 0, and the colour transition is consistent with the data obtained by the interpolation technique. Only the zones $\mathcal{Z}_{49}$ and $\mathcal{Z}_{50}$ have a risk level equal to 1, and the riskiest zones are all the zones between the zone $\mathcal{Z}_{10}$ and the zone $\mathcal{Z}_{19}$.

Concerning the second phenomenon, there can be different causes of broken rails, such as small defects in the rail, excessive loads placed on the rail, or extremely low temperatures [6]. In this analysis, we focus on this latter source of

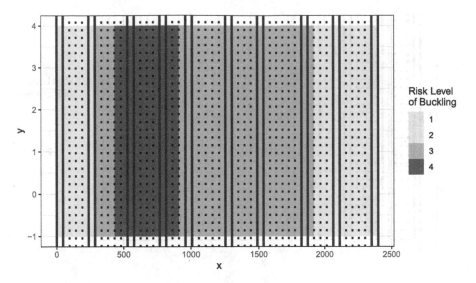

Fig. 3. Colour thermal map of the risk level of the buckling phenomenon along the considered railway line section. (Color figure online)

risk. More specifically, there exists a breaking critical value, denoted by T^*_{break}, below which the rail track temperature is so low that the breaking phenomenon can occur. According to some references, this threshold value is equal to $T^*_{\text{break}} = 10\,^{\circ}\text{C}$.

Denoted by p_{break} the probability that the breaking phenomenon occurs according to the analysis of the parameter θ, this probability can be computed as

$$p_{\text{break}} = p\big(\theta < T^*_{\text{break}}\big)\,.$$

Similarly to the previous case, given the examined phenomenon, we assign

$$\mathcal{R}_{\text{break}}(\mathcal{Z}) = \begin{cases} 0 & \text{if} & p_{\text{break}} = 0 \\ 1 & \text{if} & 0 < p_{\text{break}} \leq 0.15 \\ 2 & \text{if} & 0.15 < p_{\text{break}} \leq 0.30 \\ 3 & \text{if} & 0.30 < p_{\text{break}} \leq 0.50 \\ 4 & \text{if} & p_{\text{break}} > 0.50\,. \end{cases}$$

Figure 4 contains the colour thermal map of the risk level related to the breaking phenomenon for each zone of the area $\mathcal{A}$, obtained by using the above-described criterion. In this case, there are zones with risk levels equal to 0, i.e., all the zones between the zone $\mathcal{Z}_{10}$ and the zone $\mathcal{Z}_{19}$, which are the same zones characterised by a risk level equal to 4 for the other phenomenon. Even in this circumstance, the colour transition is consistent with the temperature estimates provided via kriging. It should be noted that there are no zones with risk levels equal to 4 for this phenomenon.

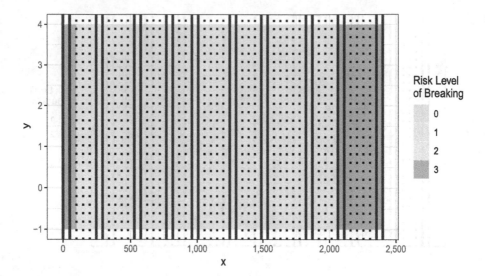

Fig. 4. Colour thermal map of the risk level of the breaking phenomenon along the considered railway line section. (Color figure online)

The following step consists of determining, for each zone, the maximum level of risk $\mathcal{R}^*(\mathcal{Z}) = \max\{\mathcal{R}_{\text{buck}}(\mathcal{Z}), \mathcal{R}_{\text{break}}(\mathcal{Z})\}$. In this case study, we obtain the following maximum risk levels:

- $\mathcal{R}^*(\mathcal{Z}) = 2$ for $\mathcal{Z} \in \{\mathcal{Z}_3, \mathcal{Z}_4, \mathcal{Z}_5, \mathcal{Z}_6\} \cup \{\mathcal{Z}_{41}, \mathcal{Z}_{42}, \mathcal{Z}_{43}\}$;
- $\mathcal{R}^*(\mathcal{Z}) = 3$ for $\mathcal{Z} \in \{\mathcal{Z}_1, \mathcal{Z}_2\} \cup \{\mathcal{Z}_7, \mathcal{Z}_8, \mathcal{Z}_9\} \cup \{\mathcal{Z}_{20}, \ldots, \mathcal{Z}_{40}\} \cup \{\mathcal{Z}_{44}, \ldots, \mathcal{Z}_{50}\}$;
- $\mathcal{R}^*(\mathcal{Z}) = 4$ for $\mathcal{Z} \in \{\mathcal{Z}_{10}, \ldots, \mathcal{Z}_{19}\}$.

4.3 Selection Criteria and Results

The last step of the risk-cost analysis involves identifying the zones that will constitute the new network $\mathcal{N}^*$.

We assume to have a budget $B = 40\,000\ €$ and a fixed cost $C = 1\,000\ €$ for each zone. With the maximum level of accuracy, each sensor has a unitary cost $C(\mathcal{R}^*(\mathcal{Z})) = C_{\mathcal{R}^*(\mathcal{Z})}$ depending on the maximum level of risk detected in the zone $\mathcal{Z}$. The Table 2 contains the unitary costs for the maximum level of risk.

In the case in which the buckling phenomenon occurs in a zone, the cost to be supported is $\mathcal{D}_{\text{buck}} = 4\,250\ €$, for each zone. On the other hand, the cost to be supported in the occurrence of a rail track breaking in a zone is $\mathcal{D}_{\text{break}} = 4\,000\ €$, for each zone. For all the zones, the cost of non-monitoring, $\mathcal{D}(\mathcal{Z})$, is computed as

$$\mathcal{D}(\mathcal{Z}) = p_{\text{buck}}(\mathcal{Z})\mathcal{D}_{\text{buck}} + p_{\text{break}}(\mathcal{Z})\mathcal{D}_{\text{break}}.$$

Table 2. Unitary costs of sensors according to the maximum risk level $\mathcal{R}^*(\mathcal{Z})$ for a zone by assuming the maximum level of accuracy.

$\mathcal{R}^*(\mathcal{Z})$	$C_{\mathcal{R}^*(\mathcal{Z})}$
0	50
1	100
2	200
3	350
4	500

Given the network $\mathcal{N}^0 = \mathcal{N}_{tot} \setminus \mathcal{N}$ and the maximum risk level in the network $\bar{\alpha} = 4$, we apply both the proposed selection criteria and compare the two different networks obtained.

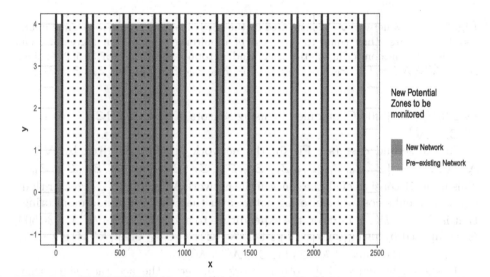

Fig. 5. The new network $\mathcal{N}^*$ constituted by the newly identified zones with a maximum risk level equal to $\bar{\alpha} = 4$ and a total expenditure of $20\,000\,\text{€}$, i.e., $\mathcal{N}^* = \bar{\mathcal{N}}^4 = \{\mathcal{Z}_{10}, \mathcal{Z}_{11}, \mathcal{Z}_{13}, \mathcal{Z}_{14}, \mathcal{Z}_{15}, \mathcal{Z}_{16}, \mathcal{Z}_{18}, \mathcal{Z}_{19}\}$.

Figures 5 and 6 show the new networks provided by selection criterion 1 and selection criterion 2, respectively. We can observe that, in both cases, the new network identified $\mathcal{N}^*$ is consistent with the risk analysis, highlighting the riskiest zones (see Figs. 3 and 4). More precisely, the first criterion (see Fig. 5) returns the network $\mathcal{N}^* = \bar{\mathcal{N}}^4 = \{\mathcal{Z}_{10}, \mathcal{Z}_{11}, \mathcal{Z}_{13}, \mathcal{Z}_{14}, \mathcal{Z}_{15}, \mathcal{Z}_{16}, \mathcal{Z}_{18}, \mathcal{Z}_{19}\}$, i.e., the network containing all the zones in the network $\mathcal{N}^0$ with the maximum level of risk, that is $\bar{\alpha} = 4$. In this case, the total expenditure to be supported is given

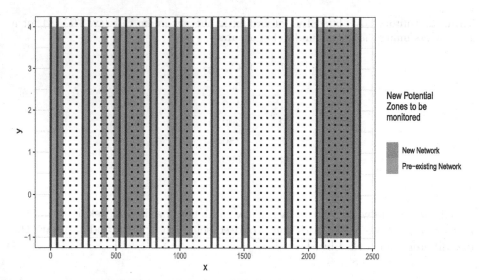

Fig. 6. The new network $\mathcal{N}^*$ constituted by the newly identified zones with a maximum risk level greater than equal to 3, a cost of monitoring the zone not greater than the cost of non-monitoring of the zone, and a total expenditure of 30 500 €, i.e., $\mathcal{N}^* = \bar{\mathcal{N}}^3 \setminus \bar{\mathcal{N}}^3_- = \{\mathcal{Z}_2, \mathcal{Z}_9, \mathcal{Z}_{11}, \mathcal{Z}_{13}, \mathcal{Z}_{14}, \mathcal{Z}_{15}, \mathcal{Z}_{20}, \mathcal{Z}_{22}, \mathcal{Z}_{23}, \mathcal{Z}_{45}, \mathcal{Z}_{46}, \mathcal{Z}_{47}, \mathcal{Z}_{48}, \mathcal{Z}_{49}\}.$

by $C^0(\bar{\mathcal{N}}^4) = 20\,000$ €, obtained by using sensors of cost $C(\mathcal{R}^*(\mathcal{Z})) = C(4)$ for all $\mathcal{Z}_i \in \bar{\mathcal{N}}^4$.

In the second case (see Fig. 6), the network provided is given by $\mathcal{N}^* = \bar{\mathcal{N}}^3 \setminus \bar{\mathcal{N}}^3_- = \{\mathcal{Z}_2, \mathcal{Z}_9, \mathcal{Z}_{11}, \mathcal{Z}_{13}, \mathcal{Z}_{14}, \mathcal{Z}_{15}, \mathcal{Z}_{20}, \mathcal{Z}_{22}, \mathcal{Z}_{23}, \mathcal{Z}_{45}, \mathcal{Z}_{46}, \mathcal{Z}_{47}, \mathcal{Z}_{48}, \mathcal{Z}_{49}\}.$ This network consists of all the zones in the network $\mathcal{N}^0$ with $\mathcal{R}^*(\mathcal{Z})$ greater than equal to 3 and a cost of monitoring not greater than the cost of non-monitoring, that is $C(\mathcal{Z}) \leq \mathcal{D}(\mathcal{Z})$. Further, the expenditure to afford is $C(\bar{\mathcal{N}}^3 \setminus \bar{\mathcal{N}}^3_-) = 30\,500$ €, computed by using sensors of cost $C(\mathcal{R}^*(\mathcal{Z})) = C(4)$ or $C(\mathcal{R}^*(\mathcal{Z})) = C(3)$ according to the value of $\mathcal{R}^*(\mathcal{Z})$ for $\mathcal{Z} \in \bar{\mathcal{N}}^3 \setminus \bar{\mathcal{N}}^3_-.$

It should be noted that both networks preserve the accuracy since they require sensors of maximum unitary cost according to the level of risk $\mathcal{R}^*(\mathcal{Z})$. The network $\mathcal{N}^* = \bar{\mathcal{N}}^4$ covers all the riskiest zones, with a total number of $|\bar{\mathcal{N}}^4| = 8$ identified zones. On the other hand, the network $\mathcal{N}^* = \bar{\mathcal{N}}^3 \setminus \bar{\mathcal{N}}^3_-$ does not contain all the riskiest zones but covers a greater number of identified zones, that is $|\bar{\mathcal{N}}^3 \setminus \bar{\mathcal{N}}^3_-| = 14$. Furthermore, in this second case, the budget surplus is less than the one generated by applying the first selection criterion.

5   Conclusions

Our investigation proposes a risk-cost analysis for tightening network monitoring in the railway infrastructure. The offered analysis can be considered as a starting point for making choices according to different decision criteria. In this regard, the monitoring and non-monitoring costs of a zone of the analysed railway line section can be combined in several ways and generate different decision-making methods.

In this work, we compare two different selection criteria for identifying and selecting additional zones of the railway line section to be controlled. Both criteria are based on the progressive removal of the zones with the lowest risk level until a total expenditure not exceeding the available budget is reached. However, if the budget constraint is not satisfied yet, the first criterion is based on the progressive reduction of the unitary cost of sensors under the same risk level. On the contrary, as long as the budget constraint does not hold, the second criterion tends to preserve the areas characterised by a monitoring cost not exceeding the non-monitoring cost.

The risk-cost analysis has been validated through a case study using simulated daily rail track temperature data. Data are acquired by 30 sensors placed along a tangent-track railway section during one calendar year. By applying the two different selection criteria, it can be noted that the second decision criterion provides an additional network containing a higher number of zones and a smaller budget surplus than the network obtained with the first criterion.

It might be interesting to analyse the results of further decision criteria, especially to find a criterion that allows us to obtain a not significant budget surplus. Furthermore, it would be possible to discuss how the network obtained with a given criterion affects the environment and identify, among some criteria, which one provides the network that guarantees a better environmental impact.

Acknowledgments. This work was partially supported by Rete Ferroviaria Italiana (RFI), the Manager of the Italian Railway Infrastructure, at which the corresponding author spent a research period as part of the project *"PON Ricerca e Innovazione" 2014–2020, Azione IV.4 "Dottorati e Contratti di ricerca su tematiche dell'innovazione"* for developing research activities about innovation, founded by the Ministero dell'Università e della Ricerca (MUR).

Appendix

Universal Kriging Interpolation Technique

The Kriging interpolation method allows for estimating the value of a variable at an unsampled location by comparing it to the values of the variable at nearby locations that have been sampled. For each structural parameter θ_j, $j = 1, \ldots, n_\Theta$, the interpolation technique is used to estimate the values of θ_j in each not monitored point identified by $P_j^* \in \mathcal{A}$.

According to the Kriging technique, the estimate of the parameter θ_j in P_j^* can be computed as a weighted average, as follows,

$$\hat{\theta}_j(P_j^*) = \sum_{i=1}^{m} \sum_{k=1}^{n_{\theta_j}} \lambda_{i,k}\, \theta_j(P_{k,j}^i), \tag{1}$$

where $\lambda_{i,k}$ are unknown weights that, under the stationarity assumption, are subject to

$$\sum_{i=1}^{m}\sum_{k=1}^{n_{\theta_j}}\lambda_{i,k}=1. \tag{2}$$

Since Kriging is designed to provide the most accurate estimate of the parameter θ_j, the weights $\lambda_{i,k}$ are chosen in order to minimise the estimate variance $\hat{\sigma}_j^2(P_j^*)$, that is,

$$\hat{\sigma}_j^2(P_j^*) = \mathrm{E}[(\theta_j(P_j^*) - \sum_{i=1}^{m}\sum_{k=1}^{n_{\theta_j}}\lambda_{i,k}\theta_j(P_{k,j}^i))^2]. \tag{3}$$

Let a point $\tilde{P}_j \in \mathcal{A}$ and a fixed distance $\vec{h}$ be considered. For a given structural parameter θ_j, the variogram function $\delta_j(\vec{h})$ is given by the following semivariance,

$$\delta_j(\vec{h}) = \frac{1}{2}\hat{\sigma}^2\left(\theta_j(\tilde{P}_j + \vec{h}) - \theta_j(\tilde{P}_j)\right),$$

depending on merely the distance $\vec{h}$. As proven in [5], the estimate variance $\hat{\sigma}_j^2(P_j^*)$ can be rewritten as a function of the variogram $\delta_j(\hat{h})$, where $\hat{h}$ is the estimated distance between two points, that is,

$$\hat{\sigma}_j^2(P_j^*) = 2\sum_{i=1}^{m}\sum_{k=1}^{n_{\theta_j}}\delta_j(\hat{h}_{*,ik}) - \sum_{i=1}^{m}\sum_{k=1}^{n_{\theta_j}}\sum_{\tilde{i}=1}^{m}\sum_{\tilde{k}=1}^{n_{\theta_j}}\lambda_{i,k}\,\lambda_{\tilde{i},\tilde{k}}\,\delta_j(\hat{h}_{ik,\tilde{i}\tilde{k}}), \tag{4}$$

where $\delta_j(\hat{h}_{*,ik})$ is the variogram depending on the distance $\hat{h}_{*,ik}$ between the points P_j^* and $P_{k,j}^i$, and similarly $\delta_j(\hat{h}_{ik,\tilde{i}\tilde{k}})$ is the variogram depending on the distance $\hat{h}_{ik,\tilde{i}\tilde{k}}$ between the points $P_{k,j}^i$ and $P_{\tilde{k},j}^{\tilde{i}}$.

The problem of minimising the estimation variance $\hat{\sigma}_j^2(P_j^*)$ in (3) is subjected to two constraints. The first one is related to the condition in (2). A further constraint is required to take into account the non-stationary trend of the parameter θ_j. In this case, we can assume that the mean of the parameter θ_j evaluated at point P_j^*, $M(P_j^*)$, is described by a polynomial function of the following type,

$$M(P_j^*) = \sum_{w=1}^{W} a_w g_w(P_j^*), \tag{5}$$

where $g_w(P_j^*)$, $w = 1, \ldots, W$, are polynomial functions of a certain order. Since Kriging is an exact estimator, by using the (1) property, expression (5) can be written as a combination of the same coefficients $\lambda_{i,k}$, that is,

$$M(P_j^*) = \sum_{i=1}^{m}\sum_{k=1}^{n_{\theta_j}}\lambda_{i,k}\mathrm{E}[\theta_j(P_{k,j}^i)] = \sum_{i=1}^{m}\sum_{k=1}^{n_{\theta_j}}\lambda_{i,k}M(P_{k,j}^i). \tag{6}$$

Combining the two Eqs. (5) and (6), the following statement holds,

$$\sum_{i=1}^{m}\sum_{k=1}^{n_{\theta_j}} \lambda_{i,k}\left(\sum_{w=1}^{W} a_w g_w(P_{k,j}^i)\right) = \sum_{w=1}^{W} a_w \left(\sum_{i=1}^{m}\sum_{k=1}^{n_{\theta_j}} \lambda_{i,k}g_w(P_{k,j}^i)\right), \qquad (7)$$

from which the second constraint of the optimization problem results

$$\sum_{i=1}^{m}\sum_{k=1}^{n_{\theta_j}} \lambda_{i,k}\, g_w(P_{k,j}^i) = g_w(P_j^*). \qquad (8)$$

Therefore, for each parameter θ_j, the optimization problem to be solved can be formulated in the following way,

$$\min_{\lambda_{i,k}} \hat{\sigma}_j^2(P_j^*) = \min_{\lambda_{i,k}}\left(2\sum_{i=1}^{m}\sum_{k=1}^{n_{\theta_j}} \delta_j(\widehat{h}_{*,ik}) - \sum_{i=1}^{m}\sum_{k=1}^{n_{\theta_j}}\sum_{\tilde{i}=1}^{m}\sum_{\tilde{k}=1}^{n_{\theta_j}} \lambda_{i,k}\,\lambda_{\tilde{i},\tilde{k}}\,\delta_j(\widehat{h}_{ik,\tilde{i}\tilde{k}})\right)$$

$$\text{s. t.}\begin{cases} \sum_{i=1}^{m}\sum_{k=1}^{n_{\theta_j}}\lambda_{i,k} = 1 \\[2mm] \sum_{i=1}^{m}\sum_{k=1}^{n_{\theta_j}}\lambda_{i,k}\,g_w(P_{k,j}^i) = g_w(P_j^*) \qquad \forall\, w = 1,\dots,W. \end{cases}$$
$$(9)$$

The optimization problem in (9) requires the Lagrangian approach. Assuming that $(\boldsymbol{\lambda},\boldsymbol{\mu}) = (\lambda_{1,1},\dots,\lambda_{m,n_{\theta_j}},\mu_0,\mu_1,\dots,\mu_W)$, where $\boldsymbol{\mu}$ is the lagrangian multipliers vector, the problem is solved by deriving the first order conditions of the following Lagrangian function,

$$\mathcal{L}(\boldsymbol{\lambda},\boldsymbol{\mu}) = \hat{\sigma}_j^2(P_j^*) - \mu_0\left(1 - \sum_{i=1}^{m}\sum_{k=1}^{n_{\theta_j}}\lambda_{i,k}\right) - \sum_{w=1}^{W}\mu_w\left(g_w(P_j^*) - \sum_{i=1}^{m}\sum_{k=1}^{n_{\theta_j}}\lambda_{i,k}\,g_w(P_{k,j}^i)\right).$$
$$(10)$$

An example of a graphical representation of the metamodel built via the Kriging interpolation is shown in Fig. 7.

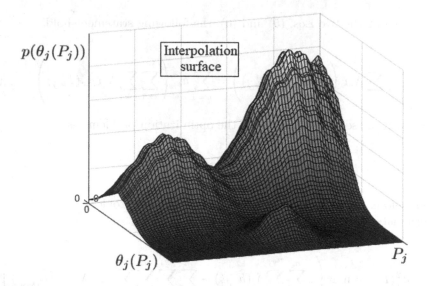

$p(\theta_j(P_j))$

$\theta_j(P_j)$

P_j

Fig. 7. Interpolation surface obtained via the Universal Kriging technique.

References

1. Bagajewicz, M., Sanchez, M.: Cost-optimal design of reliable sensor networks. Comput. Chem. Eng. **23**(11–12), 1757–1762 (2000). https://doi.org/10.1016/S0098-1354(99)00324-5
2. Balageas, D., Fritzen, C.P., Güemes, A. (eds.): Structural Health Monitoring, vol. 90. Wiley, Hoboken (2010). https://doi.org/10.1002/9780470612071
3. Capellari, G., Chatzi, E., Mariani, S.: Cost-benefit optimization of structural health monitoring sensor networks. Sensors **18**(7), 2174 (2018). https://doi.org/10.3390/s18072174
4. Chisari, C., Macorini, L., Amadio, C., Izzuddin, B.A.: Optimal sensor placement for structural parameter identification. Struct. Multidiscip. Optim. **55**, 647–662 (2017). https://doi.org/10.1007/s00158-016-1531-1
5. Cressie, N.: Statistics for Spatial Data. Wiley, Hoboken (2015). https://doi.org/10.1002/9781119115151
6. De Iorio, A., Grasso, M., Penta, F., Pucillo, G.P.: A three-parameter model for fatigue crack growth data analysis. Frattura Integr. Strutturale **6**(21), 21–29 (2012). https://doi.org/10.3221/IGF-ESIS.21.03
7. Dick, C.T., Barkan, C.P., Chapman, E.R., Stehly, M.P.: Multivariate statistical model for predicting occurrence and location of broken rails. Transp. Res. Rec. **1825**(1), 48–55 (2003). https://doi.org/10.3141/1825-07
8. Doostmohammadian, M., Rabiee, H.R., Khan, U.A.: Structural cost-optimal design of sensor networks for distributed estimation. IEEE Signal Process. Lett. **25**(6), 793–797 (2018). https://doi.org/10.1109/LSP.2018.2824761
9. Friedman, C.A., Sandow, S.: Learning probabilistic models: an expected utility maximization approach. J. Mach. Learn. Res. **4**, 257–291 (2003). https://doi.org/10.1162/153244304773633816
10. Kleijnen, J.P.: Kriging metamodeling in simulation: a review. Eur. J. Oper. Res. **192**(3), 707–716 (2009). https://doi.org/10.1016/j.ejor.2007.10.013

11. Ly, S., Charles, C., Degre, A.: Geostatistical interpolation of daily rainfall at catchment scale: the use of several variogram models in the Ourthe and Ambleve catchments, Belgium. Hydrol. Earth Syst. Sci. **15**(7), 2259–2274 (2011). https://doi.org/10.5194/hess-15-2259-2011

12. Matheron, G.: Principles of geostatistics. Econ. Geol. **58**(8), 1246–1266 (1963). https://doi.org/10.2113/gsecongeo.58.8.1246

13. Ngamkhanong, C., Kaewunruen, S., Costa, B.J.A.: State-of-the-art review of railway track resilience monitoring. Infrastructures **3**(1), 3 (2018). https://doi.org/10.3390/infrastructures3010003

14. Palin, E.J., Stipanovic Oslakovic, I., Gavin, K., Quinn, A.: Implications of climate change for railway infrastructure. Wiley Interdiscip. Rev.: Clim. Change **12**(5), e728 (2021). https://doi.org/10.1002/wcc.728

15. Pucillo, G.P.: Thermal buckling and post-buckling behaviour of continuous welded rail track. Veh. Syst. Dyn. **54**(12), 1785–1807 (2016). https://doi.org/10.1080/00423114.2016.1237665

16. Pucillo, G.P.: On the effects of multiple railway track alignment defects on the CWR thermal buckling. In: ASME/IEEE Joint Rail Conference, vol. 50978, p. V001T01A018. American Society of Mechanical Engineers (2018). https://doi.org/10.1115/JRC2018-6205

17. Pucillo, G.P.: Thermal buckling in CWR tracks: critical aspects of experimental techniques for lateral track resistance evaluation. In: ASME/IEEE Joint Rail Conference, vol. 83587, p. V001T08A009. American Society of Mechanical Engineers (2020). https://doi.org/10.1115/JRC2020-8079

18. Pucillo, G.P.: The effects of the cold expansion degree on the fatigue crack growth rate in rail steel. Int. J. Fatigue **164**, 107130 (2022). https://doi.org/10.1016/j.ijfatigue.2022.107130

19. Pucillo, G.P., Carrabs, A., Cuomo, S., Elliott, A., Meo, M.: Cold expansion of rail-end-bolt holes: finite element predictions and experimental validation by DIC and strain gauges. Int. J. Fatigue **149**, 106275 (2021). https://doi.org/10.1016/j.ijfatigue.2021.106275

20. Pucillo, G. P., De Iorio, A., Rossi, S., Testa, M.: On the effects of the USP on the lateral resistance of ballasted railway tracks. In: ASME Joint Rail Conference, JRC 2018, Pittsburgh, PA, USA, Paper n. JRC2018-6204 (2018). https://doi.org/10.1115/JRC2018-6204

21. Pucillo, G.P., Esposito, L., Leonetti, D.: On the effects of unilateral boundary conditions on the crack growth rate under cycling bending loads. Int. J. Fatigue **124**, 245–252 (2019). https://doi.org/10.1016/j.ijfatigue.2019.02.038

22. Richardson, D., et al.: Global increase in wildfire potential from compound fire weather and drought. NPJ Clim. Atmos. Sci. **5**(1), 23 (2022). https://doi.org/10.1109/MVT.2014.2333764

23. Thurston, D.F.: Broken rail detection: practical application of new technology or risk mitigation approaches. IEEE Veh. Technol. Mag. **9**(3), 80–85 (2014). https://doi.org/10.1109/MVT.2014.2333764

24. Zangl, H., Bretterklieber, T., Brasseur, G.: A feasibility study on autonomous online condition monitoring of high-voltage overhead power lines. IEEE Trans. Instrum. Meas. **58**(5), 1789–1796 (2009). https://doi.org/10.1109/TIM.2009.2012943

A Collaborative Multi-objective Approach for Clustering Task Based on Distance Measures and Clustering Validity Indices

Beatriz Flamia Azevedo[1,2,3](✉) ⓘ, Ana Maria A. C. Rocha[2]ⓘ, and Ana I. Pereira[1,2,3]ⓘ

[1] Research Centre in Digitalization and Intelligent Robotics (CeDRI), Instituto Politécnico de Bragança, 5300-253 Bragança, Portugal
{beatrizflamia,apereira}@ipb.pt
[2] ALGORITMI Research Centre/LASI, University of Minho, Campus de Gualtar, 4710-057 Braga, Portugal
arocha@dps.uminho.pt
[3] Laboratório Associado para a Sustentabilidade e Tecnologia em Regiões de Montanha (SusTEC), Instituto Politécnico de Bragança, 5300-253 Bragança, Portugal

Abstract. Clustering algorithm has the task of classifying a set of elements so that the elements within the same group are as similar as possible and, in the same way, that the elements of different groups (clusters) are as different as possible. This paper presents the *Multi-objective Clustering Algorithm* (MCA) combined with the NSGA-II, based on two intra- and three inter-clustering measures, combined 2-to-2, to define the optimal number of clusters and classify the elements among these clusters. As the NSGA-II is a multi-objective algorithm, the results are presented as a Pareto front in terms of the two measures considered in the objective functions. Moreover, a procedure named *Cluster Collaborative Indices Procedure* (CCIP) is proposed, which aims to analyze and compare the Pareto front solutions generated by different criteria (Elbow, Davies-Bouldin, Calinski-Harabasz, CS, and Dumn indices) in a collaborative way. The most appropriate solution is suggested for the decision-maker to support their final choice, considering all solutions provided by the measured combination. The methodology was tested in a benchmark dataset and also in a real dataset, and in both cases, the results were satisfactory to define the optimal number of clusters and to classify the elements of the dataset.

Keywords: clustering validity indices · multi-objective · classification

This work has been supported by FCT Fundação para a Ciência e Tecnologia within the R&D Units Project Scope UIDB/00319/2020, UIDB/05757/2020, UIDP/05757/2020 and Erasmus Plus KA2 within the project 2021-1-PT01-KA220-HED-000023288. Beatriz Flamia Azevedo is supported by FCT Grant Reference SFRH/BD/07427/2021.

1   Introduction

Clustering is one of the most widely used methods for unsupervised learning. Its main purpose is to divide the elements of a dataset into groups (clusters) based on the similarities and dissimilarities of the elements. A good clustering algorithm should maintain high similarity within the cluster and higher dissimilarities in distinct clusters. Most current clustering methods have also been proposed for integrating different distance measures to achieve the optimum clustering division. However, the weights for various distance measures are challenging to set [15]. So, a multi-objective optimization algorithm is a suitable strategy for this problem. Besides, in many cases, the estimation of the number of clusters is difficult to predict due to a lack of domain knowledge of the problem, clusters differentiation in terms of shape, size, and density, and when clusters are overlapping in nature [9]. Thus, providing a set of optimal solutions (multi-objective approach) instead of a single one (single-objective approach) is more effective, mainly in problems where human knowledge (decision-maker) is essential.

The advantage of using multi-objective strategies in the clustering task is to combine multiple objectives in parallel. In this way, it is possible to consider different distance measures and cluster quality parameters to provide a more robust and flexible algorithm. Thus, some research deeply explored these advantages in recent years. Kaur et al. [14] explore compactness and connectedness clustering properties through a multi-objective clustering algorithm based on vibrating particle system; Nayak et al. [17] present a multi-objective clustering combined with the Differential Evolution algorithm, based on three objectives related to closeness and separation between the cluster elements and also minimization of the number of the clusters; Liu et al. [15] present two multi-objective clustering approaches based on the combination of multiple distance measures; Dutta et al. [9] proposes a Multi-Objective Genetic Algorithm for automatic clustering, considering numeric and categorical features, that take advantage of the local search ability of k-means with the global search ability of MOGA to find the optimum k, intending to minimize the intra-cluster distance and maximize the inter-cluster distance. All of these presented approaches promise results for classifying elements of different datasets.

The approach proposed in this work explores different clustering measures (two intra- and three inter-clustering measures), combined 2-to-2, to develop a flexible and robust multi-objective clustering algorithm, not dependent on the initial definition of the number of centroids. For this, a Multi-objective Clustering Algorithm (MCA) was developed combined with the Non-dominated Sorting Genetic Algorithm II (NSGA-II) [6], with two intra- and three inter-clustering measures in parallel, minimizing the intra-clustering measure and maximizing the inter-clustering measure. For the six possible combinations, a Pareto front was generated, and the solutions were evaluated by five clustering validity indices (CVIs): Elbow (EI), Davies-Bouldin (BD), Calinski-Haranasz (CH), CS, and Dumn (DI) indices, through the Cluster Collaborative Indices Procedure (CCIP). This evaluation aims to refine the Pareto front solutions and support the decision-makers final choice based on different metrics proposed by each CVIs, collaboratively. The collaborative algorithm is very helpful in case the decision-maker does not know enough to select one solution from the Pareto front set since the method can suggest the most appropriate solution among the ones that belong to the Pareto front sets.

This paper is organized as follows. After the introduction, Sect. 2 describes the clustering measures, which are divided into intra- and inter-clustering measures. After that, Sect. 3 presents the clustering validity indices (CVIs). Section 4 presents the algorithm developed, the Clustering Multi-objective Algorithm (MCA), and the Cluster Collaborative Index Procedure (CCIP). The results and discussions are presented in Sect. 5. Finally, Sect. 6 presents the conclusion and future steps.

2   Clustering Measures

To classify the elements of the dataset into different groups, it is necessary to establish some measures for computing the distances between elements. The choice of distance measures is fundamental to the algorithm's performance since it strongly influences the clustering results. In this work, different clustering measures are considered to automatically define the optimal number of clusters, minimizing the intra-cluster distance and simultaneously maximizing the inter-cluster distance in a multi-objective approach.

Consider a dataset $X = \{x_1, x_2, ..., x_m\}$, where each observation is a $|d|$ - dimensional real vector. The clustering algorithm consists of partitioned the elements of X into k subsets, it is clusters, in which each cluster set is defined as $C_j = \{x_1^j, x_2^j, ..., x_i^j\}$ with $j = \{1, ..., k\}$, in other words, x_i^j represents an element i that belongs to cluster j and, on the other hand, x_l^t represents another element l that belongs to cluster t. Following, Sect. 2.1 and Sect. 2.2 present the intra- and inter-clustering measures considered, respectively.

2.1   Intra-clustering Measures

Intra-clustering measures refer to the distance among elements of a given cluster. There are many forms to compute the intra-clustering measure. Based on previous studies [3], two of them are explored in this paper, as presented bellow:

- **SMxc**: mean distances between the elements belonging to cluster C_j until its centroids, c_j.
- **FNc**: sum of the furthest neighbor distance of each cluster c_j, where x_i^j and x_l^j belong to the same cluster c_j.

2.2 Inter-clustering Measure

In turn, inter-clustering measures define the distance between elements that belong to different clusters or about the distance between different centroids c_j. In this case, three inter-clustering measures were considered [3]:

- **Mcc:** mean of the distance of all centroids.
- **MFNcc:** mean of the distances of the furthest neighbors among the different clusters, in terms of the number of clusters.
- **MNNcc:** mean of the nearest neighbor distance between elements of the different clusters.

3 Cluster Validity Indices

Cluster Validity Indices (CVIs) define a relation between intra-cluster cohesion and inter-cluster separation to assess the clustering separation quality. A CVI is expected to be able to distinguish between superior and inferior potential solutions, to guarantee the efficiency of the clustering algorithm [13]. The CVI outcome depends only on the partition provided by the clustering algorithm given a specific number of groups [10]. An optimal solution for one specific CVI could not be the optimal solution for another CVI, since each of them has shortcomings and biases [12]. In this way, there are several CVIs available in the literature, as well as several comparative studies between them, as can be seen in [1,10]. For this reason, in this work, it was chosen to use multiple CVIs, through a collaborative strategy, to reduce their shortcomings and biases. Thus, five of them were chosen, the classical ones according to the literature, and they are described below.

3.1 Elbow Index

To use the Elbow index (EI) it is necessary to evaluate the Within-Cluster Sum of Square (WCSS), which means the sum of the Euclidean distance between the elements to their centroids j, for each cluster, given by the Eq. (1). Therefore, the WCSS is the sum of all individual $WCSS_j$. When the number of clusters k is less than the optimal number of clusters, WCSS should be high, and when it increases, WCSS will follow an exponential decay. At some point, the decay will become almost linear and WCSS will continue to fall smoothly. The first point that deviates from the exponential curve is considered the elbow, and the associated number of clusters is selected as the optimum. A simplified graphic approximation to find the elbow is to draw a straight line between the WCSS values of the first (with $k = k_{min}$) and last ($k = k_{max}$) cluster solutions and calculate the distance between all the points on the curve and the straight line. Thence, the elbow is the point with the highest distance to the line [7].

$$WCSS_j = \sum_{i=1}^{\#C_j} D(x_i^j, c_j) \tag{1}$$

3.2   Davies-Bouldin Index

The Davies-Bouldin index (DB) [1], estimates the cohesion based on the distance from the elements x_i^j in a cluster to its centroid c_j and the separation based on the distance between centroids $D(c_j, c_t)$. First, it is necessary to evaluate an intra-cluster measure represented by the mean distance between each element within the cluster x_i^j and its centroid c_j, which is a dispersion parameter $S(c_k)$, as Eq. (2),

$$S(c_j) = \sum_{i=1}^{\#C_j} \frac{D(x_i^j, c_j)}{\#C_j} \tag{2}$$

in which $D(x_i^j, c_j)$ is the Euclidean distance between an element x_i^j, that belong to the cluster j, and its centroid c_j. Thus, the DB index is given by Eq. (3), where $D(c_j, c_t)$ is the Euclidean distance between the centroid c_j, and the centroid c_t, and the k is the number of clusters. The smallest DB indicates the optimal partition.

$$DB = \frac{1}{k} \sum_{j=1}^{k} \max_{t=1,\dots k, j \neq t} \left\{ \frac{S(c_j) + S(c_t)}{D(c_j, c_t)} \right\} \tag{3}$$

3.3   Calinski-Harabasz Index

The Calinski-Harabasz (CH) [4] is a ratio-type index where the cohesion is estimated based on the distance from the elements in a cluster to its centroid [1,4]. First, it is necessary to calculate the inter-cluster dispersion (BGSS), which measures the weighted sum of squared distance between the centroids of a cluster, c_j, and the centroid of the whole dataset, denoted as $\overline{X}$, which represents the barycenter of the X dataset. The BGSS is defined as Eq. (4)

$$BGSS = \sum_{j=1}^{k} \#C_j \times D(c_j, \overline{X}) \tag{4}$$

The second step is to calculate the intra-cluster dispersion for each cluster j, also given by the sum all individual within group sums of squares, $WCSS$, as defined in Eq. (1). Thus, the CH index is defined as Eq. (5):

$$CH = \frac{\#X - k}{k - 1} \times \frac{BGSS}{WCSS} \tag{5}$$

3.4   CS Index

The CS index [5] is a ratio-type index that estimates the cohesion by the cluster diameters and the separation by the nearest neighbor distance. This measure is a function of the ratio of the sum of within-cluster scatter to between-cluster separation. The smallest CS, defined by Eq. (6) indicates a valid optimal partition [5].

$$CS = \frac{\sum_{j=1}^{k}\{\frac{1}{\#C_j}\sum_{x_i^j \in \#C_j}\max_{x_l^j \in C_j}\{D(x_i^j, x_l^j)\}\}}{\sum_{j=1}^{k}\{\min_{t \in 1:k, t \neq j}\{D(c_j, c_t)\}\}} \tag{6}$$

3.5 Dumn Index

The Dumn index (DI) [8] is a ratio-type index where the cohesion is estimated by the nearest neighbor distance and the separation by the maximum cluster diameter. Thus, a higher DI will indicate compact, well-separated clusters, while a lower index will indicate less compact or less well-separated clusters [8]. So, DI is defined as the rate between the minimum distance between elements of different clusters, it is x_i^j and x_l^t, and the largest distance between elements of the same cluster, it is x_i^j and x_l^j (sometimes called cluster diameter), as defined in Eq. (7).

$$DI = \frac{\min_{j,t \in 1:k}\{D(x_i^j, x_l^t)\}}{\max_{j=1:k}\{D(x_i^j, x_l^j)\}} \tag{7}$$

4   Proposed Algorithms

This section presents the *Multi-objective Clustering Algorithm* (MCA) that, together with the NSGA-II, consists of evaluating intra- and inter-clustering measures to define the optimal number of cluster partitions (centroids) and their optimal position, minimizing the intra-cluster distance and maximizing the inter-cluster distance. As we are considering six pairs of measures, the results of the approach are six Pareto fronts, one from each pair of solutions. Furthermore, a procedure denoted as *Cluster Collaborative Indices Procedure* (CCIP) is proposed, which aims to combine and refine the Pareto front solutions using different CVI criteria, in a collaborative way. Thence, the most appropriate solution, according to all CVIs, is selected to support the decision-maker's final choice.

4.1   Multi-objective Clustering Algorithm

To explain the MCA, consider the dataset $X = \{x_1, x_2, ..., x_m\}$ composed of m elements which are intended to partition X into k groups (clusters). As the MCA can automatically define the optimal number of cluster partitions, it is necessary to define the range of possible partitions; it is the minimum and maximum number of centroids k. So, it was defined k_{min} as the minimum number of centroids, and $k_{max} = \lceil\sqrt{m}\rceil$ the maximum number of clusters that the dataset can be partitioned, where k_{max} corresponds to the integer value of the square root of the number of elements in the dataset X.

Next, the MCA randomly generates k_{max} ordered vector belonging to the domain of X, which are the possible candidates for the centroids. For each candidate, a random value ω belonging to $[0, 1]$ is associated, which will be used to select the centroids based on a threshold value γ. The centroids candidates that satisfy the constraint $\omega > \gamma$ advance to the next selection phase.

Following, the Euclidean distance between all elements from X to all centroid candidates k is evaluated. The elements closest to each centroid k define a cluster set C. To avoid small clusters sets, the centroids k that have less than α associated elements, in which $\alpha = \lceil\sqrt{m}\rceil$, are removed from the set of centroids and the elements become part of one remaining centroid, which is the closest one in terms of Euclidean distance of the elements. The remaining centroids are denoted as the centroid of each subset c_k, in which X is partitioned.

After all elements are associated with a centroid c_j, a position must be set at each coordinate to improve the performance of the algorithm. Thus, the coordinates of each centroid assume the coordinates of its cluster barycenter, composed of its elements, x_i^j.

Next, the objective functions values f_h of the problem are calculated, for $h = 2$, where f_1 represents an intra-clustering measure, chosen among the ones presented at Sect. 2.1 and f_2 represents an inter-clustering measure, chosen among the ones presented at Sect. 2.2. Therefore, The NSGA-II algorithm [6] was used to define the set of optimal solutions to the problem, that is, to define finding the Pareto front. By default, the NSGA-II is a minimization algorithm, so the f_2 values are considering negative, respecting the principle of $min\ f_2 = -max\ f_2$ [6].

4.2   Cluster Collaborative Indices Procedure

To evaluate the quality of the solutions of the Pareto fronts generated, the Cluster Collaborative Indices Procedure (CCIP) was developed. In this procedure, each solution of each Pareto front was evaluated by each CVIs criterion. As previously said, an optimal solution for one specific CVI could not be optimal for another CVI [12]. So, after evaluating each solution according to each CVI, the CCIP selects the β best solutions according to each CVIs criterion. Next, the intersection solutions between each Pareto front are evaluated, that is $FS = PF_1 \cap PF_2 \cap PF_3...PF_b$, where b is the maximum number of Pareto fronts to be evaluated, and FS defines the set of final optimal solutions composed of the solution provided by the different Pareto fronts. In this way, the solutions defined in FS are among the best β of each pair of clustering measures combination, according to all five CVIs. After that, each CVI indicates its best solution from the remaining set FS to assist the decision-maker in the solution selection. The solution with the most indications is considered the most appropriate to be selected. In case of a tie, the set of solutions indicated is considered the most appropriate for the problem, and it is up to the decision maker to take the final decision.

5   Results and Discussion

To evaluated the approaches proposed, two datasets are considered. For both dataset, the MCA parameters used were $k_{min} = 2$, $k_{max} = [\sqrt{m}]$, $\gamma = 0.4$. Since MCA is a stochastic algorithm, 10 runs was considered for each measure combination. Regarding NSGA-II parameters, a population equal to 100, maximum generation equal to $200 \times n_f$ were used, where n_f is the number of features, as default [16]. For NSGA-II, the algorithm stops when the geometric mean of the relative change in spread value over 100 generations is less than 10^{-4}, and the final spread is less than the mean spread over the past 100 generations, as defined in `gamultiobj` function [16] documentation.

5.1   Results from Dataset 1

The dataset 1 is a benchmark dataset composed of 300 elements and 2 attributes, available at [11], which indicates 3 clusters as the optimal solution by a single objective approach. Thus, the two intra-clustering measures ($SMxc$ and FNc) are combined with the three inter-clustering measures (Mcc, $MFNcc$, and $MNNcc$), 2-to-2, with the first objective function being an intra-clustering measure, and the second objective function being an inter-clustering measure. This combination results in six Pareto fronts, but with different ranges since they involve measures of sums and means. Thus, to have a fair comparison between the Pareto fronts, they have been normalized. The results of this manipulation are presented in Fig. 1a and Fig. 1b illustrating the same six Pareto fronts, but also showing the number of clusters k on the z-axis.

Note that, according to the algorithm parameter, the maximum number of k allowed is 17. However, in the model solution, 14 was the maximum number of clusters in a Pareto front, provided by the combination $SMxc$ and $MNNcc$, while for the other combinations, the maximum k is 13. As for the number of solutions, 456 solutions were found, being 91, 78, 97, 69, 61, and 60 solutions, respectively for the combinations $SMxc - Mcc$, $SMxc - MFNcc$, $SMxc - MNNxx$, $FNc - Mcc$, $FNc - MFNcc$, and $FNc - MNNcc$. It is important to highlight that all solutions presented in Fig. 1 are optimal for the problem since they all belong to a non-dominated solution set of solutions of their pair of measures.

After that, these solutions were evaluated by the five CVIs indices (EI, DB, CH, CS, and DI). Since the solution is intended to be refined according to the criteria of each CVI, $\beta = 0.75$ was defined, which means that the 75% best solutions according to each CVI are kept, and the intersection set of these remaining solutions is calculated. Figure 2 illustrates the results of the Pareto fronts. Thus, Fig. 2a illustrates the Pareto fronts result, and Fig. 2b illustrates the Pareto fronts with the number of cluster k indication on the z-axis. In this case, it

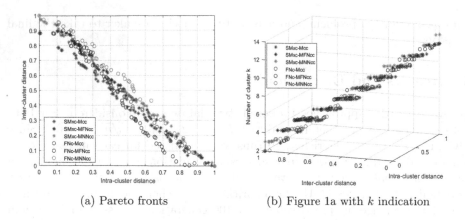

(a) Pareto fronts

(b) Figure 1a with k indication

Fig. 1. Pareto fronts of dataset 1

is possible to note that the maximum k is 9. So, all solutions with k larger than 9 were removed, as well as the other solutions that do not belong to the 75% best solutions for each CVI. Thus, the FS set is composed by 118 solutions, it is, 16 for the combinations $SMxc - Mcc$, 16 for $SMxc - MFNcc$, 37 for $SMxc - MNNxx$, 20 for $FNc - Mcc$, 14 for $FNc - MFNcc$, and 15 for $FNc - MNNcc$, which represent a 74% reduction relatively to the initial set of optimal solutions.

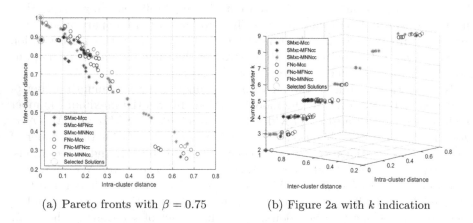

(a) Pareto fronts with $\beta = 0.75$

(b) Figure 2a with k indication

Fig. 2. Pareto fronts of dataset 1, considering $\beta = 0.75$

After that, each CVI identifies its best solution from the remaining set. Thus, the solution with the most indications is considered the most appropriate to be selected. Furthermore, in case of a tie, the set of solutions indicated is also considered the most appropriate for the problem. Considering the remaining solutions, in dataset 1, there was a tie between four solutions provided by indication of the indices: DB, CH, CS. Figure 3 illustrates these four solutions. Although they all divide the dataset into 3 sets, the centroids' position and the distribution of the elements are different.

Solution 1 and 2 are provided by the objective function 1 being the $SMxc$ and the objective function 2 being the Mcc; while solution 3 and 4 are provided by $SMxc$ and $MNNcc$, objective functions 1 and 2, respectively. Thereby, solution 1 centroids are denoted as $c_1 = (0.019, -0.032)$, $c_2 = (5.978, 1.004)$, and $c_3 = (2.713, 4.104)$, that can be analyzed in Fig. 3a. Solution 2 centroids are $c_1 = (-0.008, -0.065)$, $c_2 = (6.011, 0.977)$, and $c_3 = (2.719, 4.020)$ - Fig. 3b. Solution 3 centroids are $c_1 = (0.019, -0.032)$, $c_2 = (5.978, 1.003)$, $c_3 = (2.713, 4.105)$ - Fig. 3c. And, solution 4 centroids are defined as $c_1 = (0.019, -0.032)$, $c_2 = (5.978, 1.004)$, and $c_3 = (2.713, 4.105)$ - Fig. 3d. Thus, according to the results, there is no doubt that the most appropriate number of k is 3, which goes to the solution of [11]. Thence, it is only up to the decision-maker to choose (if necessary) the distribution of the elements for the problem or just select one of the four solutions, that are approximately equal solutions.

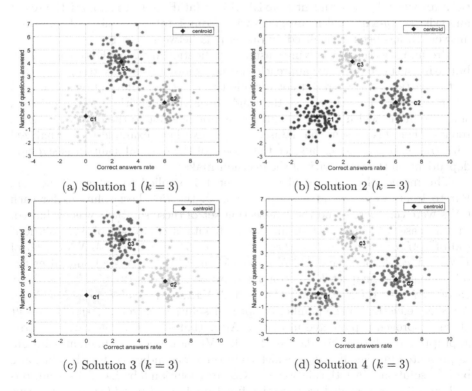

(a) Solution 1 ($k = 3$)   (b) Solution 2 ($k = 3$)

(c) Solution 3 ($k = 3$)   (d) Solution 4 ($k = 3$)

Fig. 3. Final Pareto front solutions (dataset 1)

5.2   Results from Dataset 2

To test the approach on a real case study, the methodology previously presented for dataset 1, was also applied for the dataset 2. Dataset 2 is a real case study composed of 291 elements and 2 instances, and it is provided by the MathE project [2]. The MathE project aims to provide any student all over the world with an online platform to help them to learn college mathematics and also support students who want to deepen their knowledge of a multitude of mathematical topics at their own pace. More details about the MathE project are described in [2], and can also be found on the platform Website (mathe.pixel-online.org). One of the particularities of the MathE platform is the *Student's Assessment* section, which is composed of multiple-choice questions for the students to train and practice their skills. The answers provided by each student over the 3 years that the platform has been online define the dataset 2. Therefore, each dataset element refers to one student who used the Student Assessment section. And the first instance represents the rate of the correct answer (x-axis) provided by the student's history, and the second instance represents the number of questions answered by this student (y-axis) while MathE user. To support the result analysis, the y-axis, which initially varies from 1 to 42 (number of questions answered), has been normalized by range; it is between 0 to 1.

Preliminary studies involving cluster classification and MathE students' data, but using a single objective approach, did not show satisfactory results [2]; that is, the patterns extracted did not provide the necessary information to be used by the project. This is because the single objective algorithm only provides a single solution, which, although optimal, is not relevant to the decision-maker's request. For this reason, the dataset 2 is an excellent example to be analyzed with the proposed approach since the choice of the optimal solution is strongly dependent on the sensitivity of the decision-maker.

The methodology described for dataset 1 is applied to dataset 2, i.e., six Pareto fronts were generated and normalized, the 75% best solutions for each CVIs were considered, and the intersection set of these solutions was evaluated. In this case, the initial set of all Pareto fronts is composed of 312 solutions (50 of $SMxc - Mcc$, 46 of $SMxc - MFNcc$, 54 of $SMxc - MNNxx$, 60 of $FNc - Mcc$, 48 of $FNc - MFNcc$, and 54 of $FNc - MNNcc$), and after the refinement, the final set FS is composed by 64 solutions (21 of $SMxc - Mcc$, 18 of $SMxc - MFNcc$, 7 of $SMxc - MNNxx$, and 18 of $FNc - MNNcc$). An 80% reduction in the number of optimal solutions is verified and the result of this approach is presented in Fig. 4a. After that, each CVI indicates its most appropriated solution. For dataset 2 each CVI indicated one different solution, as presented in Figs. 4b– 4f, in which solution 1 was indicated by EI, solution 2 by DB, solution 3 by CH, solution 4 by CS, and solution 5 by DI. Thus, solutions 1, 2, and 3 were provided by objective function 1 equal to $SMxc$ and objective function 2 equal to Mcc. Whereas, solutions 4 were given by objective function 1 equal to $SMxc$ and objective function 2 equal to $MNNcc$. And, solutions 5 were resulted by objective function 1 equal to FNc and objective function 2 equal to $MFNcc$. Thereby, the centroids of solution 1 are $c_1 = (0.958, 0.058)$ and $c_2 =$

$(0.391, 0.335)$, in Fig. 4b. The centroids of solution 2 are $c_1 = (0.152, 0.153)$, $c_2 = (0.636, 0.245)$, $c_3 = (1, 0.004)$, and $c_4 = (0.365, 0.705)$, as depicted in Fig. 4c. The centroids of solution 3 are $c_1 = (0.000, 0.033)$, $c_2 = (1, 0.023)$, $c_3 = (0.215, 0.321)$, $c_4 = (0.621, 0.239)$, and $c_5 = (0.546, 0.866)$, as can be seen in Fig. 4d. The 4 solution centroids are $c_1 = (0.390, 0.333)$, and $c_2 = (0.964, 0.071)$ - Fig. 4e. The centroids of solution 5 are $c_1 = (0.060, 0.120)$, $c_2 = (0.284, 0.117)$, $c_3 = (0.445, 0.111)$, $c_4 = (0.635, 0.137)$, $c_5 = (0.965, 0.057)$, $c_6 = (0.194, 0.435)$, $c_7 = (0.469, 0.366)$, $c_8 = (0.720, 0.396)$, $c_9 = (0.298, 0.777)$, and $c_{10} = (0.629, 0.831)$, in Fig. 4f.

Knowing the profile of students enrolled in the MathE platform, it is known that there is a diversity of students with different backgrounds (country, age, course and university year attending, and level of difficulty in Mathematics, among others). Therefore, a division into a few groups is not a significant result for the project, given the diversity of the public, especially in terms of performance in mathematical disciplines, as already explored in previous works. Thus, considering the previous information and interest of the MathE Project, solution 5, in Fig. 4f, is chosen as the most appropriate real one.

In solution 5, the dataset was divided into 10 clusters. In terms of the number of questions answered, clusters 1 to 5 are composed of students who answer a few questions. In contrast, clusters 6, 7, and 8 comprise students who answer a larger number of questions than the previously mentioned groups. Finally, clusters 9 and 10 are made up of students who answered the most quantity of questions on the platform. In terms of performance (correct answers rate), considering clusters 1 to 5, the students' performance increases gradually for cluster 1 to cluster 5, so in cluster 1 almost all students have a success rate equal to 0, while in cluster 5 almost all students had 1. Here it is important to point out that dataset 2 is composed of multiple equal entries (student with an equal number of questions answered and equal performance), which overlap on the graph; for this reason, cluster 5, although it seems to be composed by few students, actually includes 22 students, with 17 having 1 question answered and 1 correct answer. In clusters 6, 7, and 8 the students used the platform more than in the previous groups. In this case, the students of cluster 6 performed less than 0.35, whereas the students of cluster 7 performed between 0.35 and 0.6, and the students who performed higher than 0.6, are in cluster 8. In clusters 9 and 10, the students answered more questions. Regarding their performance, in cluster 9 they have a performance lower than 0.45, while in cluster 10 the student rate performance is higher than 0.45. In this way, the division provided by solution 5 can be used to extract valuable characteristics about the student's performance according to the group to which they belong.

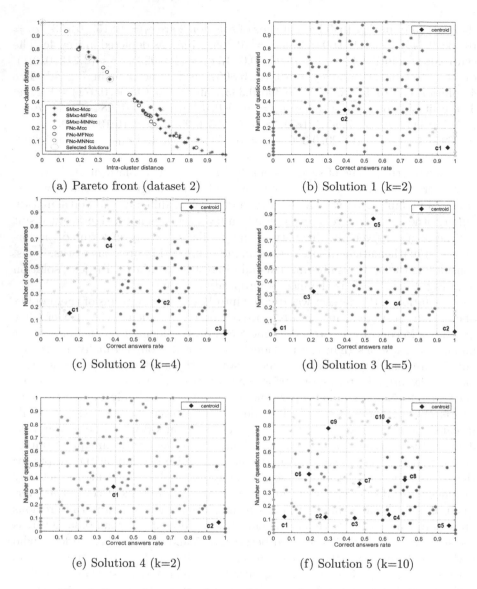

(a) Pareto front (dataset 2)

(b) Solution 1 (k=2)

(c) Solution 2 (k=4)

(d) Solution 3 (k=5)

(e) Solution 4 (k=2)

(f) Solution 5 (k=10)

Fig. 4. Pareto front and solutions (dataset 2), considering $\beta = 0.75$

6   Conclusion and Future Works

The advantage of using multi-objective strategies in the clustering task is to combine multiple objectives in parallel, such as different distance measures. This paper explored clustering measures to develop the Multi-objective Clustering Algorithm. The results of MCA consist of a set of Pareto front solutions, provided by the pair of measures, that were considered as the objective functions of a bi-objective optimization problem. The problem aimed to minimize the intra-clustering measure and maximize the inter-clustering measure using the NGSA-II. In this case, the objective function 1 refers to the intra-clustering measures, which could be the measures $SMxc$ or FNc; and the objective function 2 refers to the inter-clustering measures, which could be any one of the measures Mcc, $MFNcc$, or $MNNcc$. Besides, a procedure denoted as Cluster Collaborative Indices Procedure was proposed, aiming to compare and refine the Pareto front solutions generated by the MCA and NSGA-II, using different criteria provided by five CVIs: Elbow (EI), Davies-Bouldin (BD), Calinski-Haranasz (CH), CS, and Dumn (DI) indices. Thus, the optimal β solutions were selected according to each CVI, and the worst $(1 - \beta)$ solutions of each CVI are removed. The intersection set between each CVI β solution is calculated; finally, each CVI indicates its most appropriate solution of the intersection set. The solution with more indications is suggested to the decision-maker as the most appropriate one.

By the range and variability of each Pareto front generated, it is possible to perceive the impact of combining different measures to solve a problem. Analyzing the results of dataset 1, by Fig. 1b, it is evident that only the combination $SMxc - MFNcc$ provided solutions with $k = 14$, whereas the combination $FNc - MFNcc$ does not have solutions with k less than 4. In this way, if only one pair of measures were considered, the final solution was restricted to the optimum provided by the pair of measures combination. So, considering the results of the six Pareto fronts, the final solution is enriched by the solution provided by different measures. As already mentioned by [12], an optimal solution for one specific CVI could not be the optimal solution for another CVI due to their metrics. Considering this, choosing the most appropriate CVI for the problem is not a simple task. The intersection strategy serves to refine the solutions and ensure that all the remaining are the most appropriate β for each of the CVI, as it is very hard to achieve a solution that is the best for all CVI.

The indication of the best CVI solution is useful to help the decision-maker since even after selecting the most appropriate optimal solutions, there are still many options left, and in certain cases, the decision-maker does not have enough information about the data to quickly determine, among the set of optimal solutions, the one that most represent the problem. According to [11], considering a single objective strategy, the optimal solution for dataset 1 is 3 clusters, it is $k = 3$. As shown in Fig. 3, all the solutions indicated from dataset 1 have $k = 3$, demonstrating the effectiveness of the proposed method in a benchmark problem.

In the case of dataset 2, the data distribution is more complex than in dataset 1 [9], since the multiple points and clusters overlap, are not rounded shape, and the elements are not as well separated as the dataset 1. Thus, for dataset 2, which describes a real problem, the multi-objective strategy is much more effective than the single one since in the multi-objective, it is possible to compare and choose among a set of optimal solutions, the one that goes from meeting the patterns that the decision-maker wants to extract from the dataset. For the dataset 2, the decision maker's knowledge is of great value in defining the solution to be used.

Then, the proposed method is an asset in situations where the single objective approach is insufficient.

In the future, it is expected to explore more deeply the intra- and inter-clustering measures in multiple objective functions, as well as cluster splitting and merging strategies to improve the quality of cluster partitioning.

References

1. Arbelaitz, O., Gurrutxaga, I., Muguerza, J., Pérez, J.M., Perona, I.: An extensive comparative study of cluster validity indices. Pattern Recogn. **46**(1), 243–256 (2013)
2. Azevedo, B.F., Rocha, A.M.A.C., Fernandes, F.P., Pacheco, M.F., Pereira, A.I.: Evaluating student behaviour on the mathe platform - clustering algorithms approaches. In: Book of 16th Learning and Intelligent Optimization Conference - LION 2022, pp. 319–333. Milos - Greece (2022)
3. Azevedo, B.F., Rocha, A.M.A.C., Pereira, A.I.: A multi-objective clustering approach based on different clustering measures combinations. Submitted to Computational & Applied Mathematics
4. Caliński, T., Harabasz, J.: A dendrite method for cluster analysis. Commun. Stat. **3**(1), 1–27 (1974)
5. Chou, C.H., Su, M.C., Lai, E.: A new cluster validity measure and its application to image compression. Pattern Anal. Appl. **7**, 205–220 (2004)
6. Deb, K., Pratap, A., Agarwal, S., Meyarivan, T.: A fast and elitist multiobjective genetic algorithm: NSGA-II. IEEE Trans. Evol. Comput. **6**(2), 182–197 (2002)
7. Delgado, H., Anguera, X., Fredouille, C., Serrano, J.: Novel clustering selection criterion for fast binary key speaker diarization. In: 16th Annual Conference of the International Speech Communication Association (NTERSPEECH 2015) (2015)
8. Dunn, J.C.: A fuzzy relative of the ISODATA process and its use in detecting compact well-separated clusters. J. Cybern. **3**(3), 32–57 (1973)
9. Dutta, D., Sil, J., Dutta, P.: Automatic clustering by multi-objective genetic algorithm with numeric and categorical features. Expert Syst. Appl. **137**, 357–379 (2019)
10. Gurrutxaga, I., Muguerza, J., Arbelaitz, O., Pérez, J.M., Martín, J.I.: Towards a standard methodology to evaluate internal cluster validity indices. Pattern Recogn. Lett. **32**(3), 505–515 (2011)
11. Heris, M.K.: Evolutionary data clustering in matlab (2015). https://yarpiz.com/64/ypml101-evolutionary-clustering
12. Jain, M., Jain, M., AlSkaif, T., Dev, S.: Which internal validation indices to use while clustering electric load demand profiles? Sustain. Energy Grids Netw. **32**, 100849 (2022)
13. José-García, A., Gómez-Flores, W.: A survey of cluster validity indices for automatic data clustering using differential evolution. In: Proceedings of the Genetic and Evolutionary Computation Conference, pp. 314–322 (2021)
14. Kaur, A., Kumar, Y.: A multi-objective vibrating particle system algorithm for data clustering. Pattern Anal. Appl. **25**(1), 209–239 (2022)
15. Liu, C., Liu, J., Peng, D., Wu, C.: A general multiobjective clustering approach based on multiple distance measures. IEEE Access **6**, 41706–41719 (2018)
16. MATLAB: Mathworks inc (2019). www.mathworks.com/products/matlab.html
17. Nayak, S.K., Rout, P.K., Jagadev, A.K.: Multi-objective clustering: a kernel based approach using differential evolution. Connect. Sci. **31**(3), 294–321 (2019)

Iranian Architectural Styles Recognition Using Image Processing and Deep Learning

Mohammad Tayarani Darbandy[1]([✉]), Benyamin Zojaji[2], and Fariba Alizadeh Sani[3]

[1] School of Architecture, Islamic Azad University Taft, 8991985495 Taft, Iran
mohammad.tayarani97@gmail.com
[2] Sadjad University, Mashhad, Iran
[3] Mashhad University of Medical Science, Mashhad, Iran

Abstract. Iranian architecture, also known as Persian architecture, encompasses the design of buildings in Iran and extends to various regions in West Asia, the Caucasus, and Central Asia. With a rich history dating back at least 5,000 BC, it boasts distinctive features and styles. Iran, located in the Middle East, has faced ongoing geopolitical challenges, including the potential for conflicts, such as those in Iraq and Afghanistan. Unfortunately, historical monuments often become unintentional casualties during wartime, suffering damage or destruction. These historical monuments hold cultural and historical significance not only for the country they belong to but for all of humanity. Therefore, it is crucial to make efforts to preserve them. In this paper, we propose the development of an automated system utilizing Deep Learning methods for the detection and recognition of historical monuments. This system can be integrated into military equipment to help identify the architectural style of a building and determine its construction date. By doing so, it can provide a critical warning to prevent the targeting of historically significant structures. To support our system, we have curated a dataset consisting of approximately 3,000 photographs showcasing six distinct styles of Iranian historical architecture. Figure 1 provides some examples of these photographs. It is worth noting that this dataset can be valuable for various scientific research projects and applications beyond our proposed system. Additionally, it offers tourists the opportunity to learn about Iranian historical monuments independently, using their mobile phones to access information about a monument's historical period and architectural style, eliminating the need for a traditional guide. This initiative aims to safeguard the invaluable cultural heritage of Iran and neighboring regions, contributing to the collective preservation of these historical treasures.

Keywords: Iranian architecture · Deep learning · Image classification · Image processing

1 Introduction

The origins of Persian architecture trace back to the seventh millennium BC [1]. Iran has been at the forefront of incorporating mathematics, geometry, and astronomy into architectural practices [2]. Persian architecture is renowned for its extensive diversity in

H. Moosaei et al. (Eds.): DIS 2023, LNCS 14321, pp. 69–82, 2024.
https://doi.org/10.1007/978-3-031-50320-7_5

both structural and aesthetic aspects, evolving gradually and with precision over time. Traditionally, the guiding and formative motif of Persian architecture has been celebrated for its cosmic symbolism, serving as a means through which individuals connect and participate in the celestial realms [3]. Notably, Iran holds the seventh position globally in terms of the number of renowned landmarks and monuments listed on UNESCO's World Heritage list [4].

As noted by Arthur Pope, a prominent Persian historian and archaeologist, architecture is regarded as the foremost Iranian art, in the truest sense of the word. The prominence of architecture in Iran has been evident throughout both ancient and Islamic periods [5]. The enduring significance of Iranian architecture has resulted in challenges for individuals and students who find it difficult to distinguish between different historical periods and architectural styles of Iranian monuments. Furthermore, the task of reviewing a vast number of documents associated with each monument poses a time-consuming challenge for cultural and tourism organizations.

There are millions of images and videos available on the internet, which provide ample opportunities for advanced semantic analysis applications and algorithms. These applications and algorithms are designed to analyze images and videos [6], aiming to enhance user search experiences and provide effective summarization. Numerous researchers worldwide have reported innovative breakthroughs in fields such as image labeling, object detection, and scene classification [7, 8]. These breakthroughs have paved the way for more efficient approaches to addressing object detection and scene classification challenges. Artificial neural networks, particularly convolutional neural networks (CNNs), have shown remarkable success in implementing object detection and scene classification tasks. However, deep learning involves the analysis of vast amounts of data through neural networks, making it a complex challenge to understand and cater to user preferences online [9–11]. Torch, a popular deep-learning framework, focuses on the techniques and strategies essential for deep learning and neural networks [12, 13]. In this article, we present a novel approach to classify historical monuments using artificial intelligence and deep neural networks.

2 Background Research

The present study represents a pioneering effort in the recognition of Iranian architectural styles using deep learning techniques. In a similar vein, Jose et al. utilized Deep Learning Techniques to curate a new dataset comprising approximately 4000 images for the classification of Architectural Heritage Images [14]. Encouragingly, their study yielded promising results in terms of accuracy, suggesting that these techniques could find extensive applications in the digital documentation of architectural heritage. Husain et al. also introduced a method for designing a robust model capable of accurately detecting and classifying multi-class defects, even when working with relatively small datasets [15]. Additionally, Konstantinos et al. developed a system for the automatic detection and identification of urban buildings. They harnessed deep learning, particularly convolutional neural networks, to address this challenge, enabling the automatic generation of highly detailed features directly from source data. Their research extended to the identification of half-timbered constructed buildings in Calw city, Germany [16].

Abraham et al. incorporated sparse features alongside primary color pixel values and devised a convolutional neural network for the classification of building images. As a result, they successfully developed a neural model capable of classifying the architectural styles of Mexican buildings into three categories: prehispanic, colonial, and modern, achieving an impressive accuracy rate of 88.01% [17]. In a separate study, conducted by Aditya et al., the objective was to extract high-level image features from low-level visual features, especially when the test image belonged to one of the predefined categories. Their approach aimed to address multi-class classification by integrating two-class classifiers into a single hierarchical classifier. Their findings indicated that learning specific high-level semantic categories could be accomplished by utilizing particular low-level visual features, provided that the images were related to one of the predefined classes [18].

In a separate study, Sebastian et al. introduced a deep learning-based approach for table detection in document images. Furthermore, they introduced a novel deep learning model designed to recognize table structures, including the identification of rows, columns, and cell locations within detected tables [19]. Gaowei et al. developed a hierarchical deep learning network with a symmetric loss formula to target the identification of exterior façade elements in building images. This framework holds significant potential for enabling the automatic and precise identification of various building façade elements, particularly in complex environments. Such capabilities could prove invaluable for infrastructure monitoring and maintenance operations [20].

3 Materials and Methods

Computer vision techniques are increasingly being employed to streamline and enhance the processes involved in documenting, preserving, and restoring architectural heritage. In our work, we leverage deep learning for image classification to document architectural cultural heritage. Our model has been trained on a dataset comprising approximately 3,000 images. To maximize the effectiveness of our training, we partitioned our dataset into a training set and a test set. Given our limited data, we allocated 90% of the images for the training stage and reserved 10% of the images for testing, with 50 randomly chosen images per class. It's important to note that within the training set (90%), we further divided a portion for validation (10%). This validation subset was utilized to fine-tune our hyperparameters during the training process. Our primary objective is to develop software applications that leverage deep learning-based computer vision techniques to classify the technical knowledge used to acquire images of architectural heritage. Regardless of the method employed to acquire these images, our innovation, particularly convolutional neural networks, proves to be a valuable tool for these tasks.

In this paper, we employed Fine-tuning as a strategy to expedite the process of drawing conclusions. Fine-tuning involves the utilization of a set of pre-trained weights for training a convolutional neural network, as previously mentioned. This technique has proven effective in various applications. Given the limited amount of data available, we incorporated diverse image augmentations such as horizontal flips, random brightness shifts, shearing, and more to enhance our model's performance.

It is important to emphasize that the models we have employed have already undergone training and fine-tuning. As previously mentioned, we utilize transfer learning, which involves the use of well-established architectures, such as ResNet [21, 22], as the base classifier model, and we leverage the weights they have acquired from training on ImageNet [23–26]. A notable exception is the final fully connected layer, where the number of nodes corresponds to the classes within the dataset we intend to classify. A common practice is to replace the last fully connected layer of the pre-trained CNN [27, 28] with a new fully connected layer containing as many neurons as there are classes in the new application. Consequently, these models were initially designed for 1000 classifications, and modifications were made to the final layer of the MLP. For instance, we altered the final layer of the MLP to have 6 output features instead of the original 1000. In some models, we adjusted the output features to 64, followed by the addition of a fully connected layer that connected the 64 features to the final 6 output classes.

4   Activation Function

In this model, we utilize the SoftMax function [29–31] as the final activation function, as our objective is to categorize buildings. Since we work with PyTorch [32, 33], we also employ the Cross-Entropy Loss function [34, 35], which essentially combines the Log-Softmax [36, 37] and Negative Log-Likelihood Loss (NLLLoss) [38, 39]. To mitigate overfitting, we have incorporated dropout layers [40, 41] into the network architecture [42, 43]. For the training and inference stages, we rely on a Linux (Ubuntu) environment and utilize Nvidia 1080-Titan hardware resources.

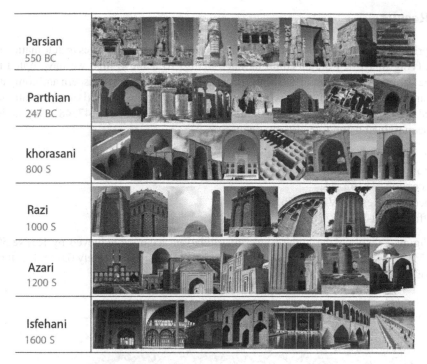

Parsian 550 BC	
khorasani 800 S	
Parthian 247 BC	
Razi 1000 S	
Azari 1200 S	
Isfehani 1600 S	

Fig. 1. Categorization of Iranian Architectural Styles

Examples of images depicting ancient Iranian architectural structures, along with information on their age and the classification of Iranian architectural styles, are illustrated in Fig. 1 from an architectural perspective. It's worth noting that all the photographs featured in this article were captured and curated by our team. Mohammad Karim Pirnia, a respected and renowned Iranian planner, categorizes the classical architecture of Iran across different historical eras into the following six styles [44]:

1. Persian Style (up to the 3rd century B.C.)
2. Parthian Style
3. Khorasani Style (from the late 7th to the late 10th century)
4. Razi Style (from the 11th century to the time of the Mongol invasion)
5. Azari Style (from the late 13th century until the emergence of the Safavid Dynasty in the 16th century)
6. Isfahani Style, which spans across the Safavid, Afsharid, Zand, and Qajar dynasties, starting from the 16th century and continuing onwards.

5 Results

The primary objective is to evaluate the effectiveness of these techniques in automatically classifying images of architectural heritage items. To achieve this, we have opted to utilize various types of neural networks, as our experiments aimed to discern and compare the suitability of different options already in use. Notably, we focused on convolutional neural networks (CNN) [45, 46] and multi-layer perceptrons (MLP) [47, 48]. Our model has been trained using five distinct architectures:

1. ResNet50 [49, 50]
2. MobilenetV3-Large [51, 52]
3. Inception-V4 [53, 54]
4. Inception-ResNet-V2 [55, 56]
5. EfficientNet-B3 [57, 58]

Figure 2 displays the highest level of validation accuracy achieved by ResNet50, with MobileNet-V3 achieving a slightly lower accuracy, approximately 0.02% less than that of ResNet50.

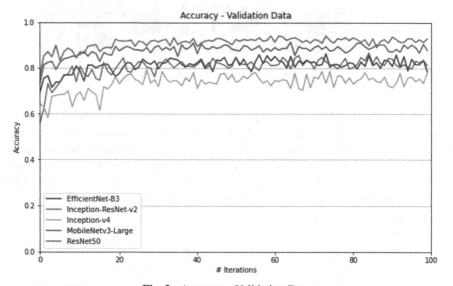

Fig. 2. Accuracy - Validation Data

Figure 3 demonstrates that using Inception-ResNet-V2, the loss count extends up to 40 epochs, surpassing even MobileNetv3-large in terms of loss. Additionally, it is noteworthy that the loss in Inception-ResNet-V2 is substantially lower compared to ResNet50, indicating a significant difference.

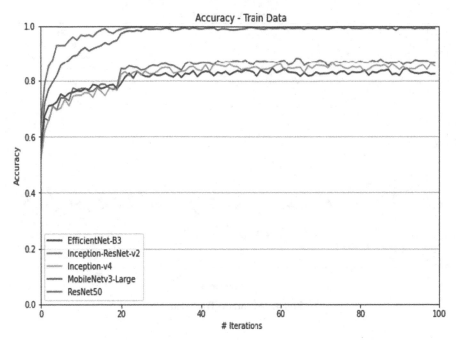

Fig. 3. Loss - Validation Data

In Fig. 4, the training accuracy is depicted, reaching its peak in ResNet50, followed closely by MobileNetv3-Large in the second position. Throughout the training process, the accuracy between Inception-v4 and Inception-Resnet-v2 remains very similar, with results being close to each other. Conversely, the lowest accuracy is observed in EfficientNet-B3. Referencing Table 1, it's evident that ResNet exhibits the highest accuracy and the lowest loss, while Inception-v4 records the lowest accuracy and the highest loss. It's important to note that the standard deviation of our dataset is 89.13.

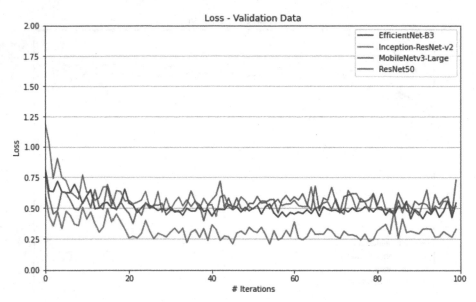

Fig. 4. Accuracy-Train data

Table 1. Compares train and test

	Train		Test	
	Loss	Accuracy	Loss	Accuracy
EfficientNet-B3	0.4793	82.50%	0.8617	83.11%
Inception-ResNet-v2	0.3848	86.47%	2.1547	51.09%
Inception-v4	0.3927	85.35%	18.7749	48.79%
MobileNetv3-Large	0.026	99.19%	0.6143	88.37%
ResNet	0.0019	99.41%	0.2794	95.06%

Furthermore, in Table 2, standard metrics for point-based classification are presented. We provide Precision, Recall, and F1-Score [59, 60] for each class, along with the mean averages of these metrics. The standard F1-Score is calculated as the harmonic mean of precision and recall. A perfect model would achieve an F-score of 1.

Table 2. Precision of Iranian styles

	Precision	Recall	F1-score	Quantity
Azari	0.96	1.00	0.98	**50**
Isfehani	0.98	0.90	0.94	**50**
Khorasani	0.87	0.96	0.91	**50**
Parsian	0.96	0.96	0.96	**50**
Parthian	0.98	0.90	0.94	**50**
Razi	0.96	0.98	0.97	**50**
Accuracy	**95**			

Table 3 presents the dataset image resolutions employed during various training phases, each spanning 100 epochs. The confusion matrix is depicted in Table 4.

Table 3. Image sizes

Size	Inference time	
EfficientNet-B3	300 * 300	0.011
Inception-ResNet-v2	299 * 299	0.006
Inception-v4	299 * 299	0.003
MobileNetv3-Large	224 * 224	0.001
ResNet50	224 * 224	0.001

Table 4. Confusion Matrix

Confusion matrix

Actual / Predicted

	Azari	Isfehani	Khorasani	Parsian	Parthian	Razi	sum_lin
Azari	50 / 16.67%						50 / 100% / 0.00%
Isfehani	1 / 0.33%	45 / 15.00%	4 / 1.33%				50 / 90.00% / 10.00%
Khorasani			48 / 16.00%			2 / 0.67%	50 / 96.00% / 4.00%
Parsian			1 / 0.33%	48 / 16.00%	1 / 0.33%		50 / 96.00% / 4.00%
Parthian	1 / 0.33%		2 / 0.67%	2 / 0.67%	45 / 15.00%		50 / 90.00% / 10.00%
Razi		1 / 0.33%				49 / 16.33%	50 / 98.00% / 2.00%
sum_col	52 / 96.15% / 3.85%	46 / 97.83% / 2.17%	55 / 87.27% / 12.73%	50 / 96.00% / 4.00%	46 / 97.83% / 2.17%	51 / 96.08% / 3.92%	300 / 95.00% / 5.00%

Predicted

6 Conclusion

Given the significance of Iranian architecture and the need to classify its various styles, it becomes essential to organize archives in several institutions, including the Tourism and Cultural Heritage Organization. This becomes particularly crucial in situations where a quick identification of architectural style and age is required. Our approach helps identify Iran's cultural heritage swiftly, even during times of conflict, to prevent damage and destruction of these valuable structures. In summary, our deep-learning system successfully categorized buildings and compared various architectural styles. We achieved an impressive accuracy of 95% with ResNet50, a result that can be considered reliable and trustworthy. It's worth noting that with a larger dataset, we anticipate achieving even higher accuracy levels. Looking ahead, our future plans involve expanding the network's capabilities to analyze closer details of building textures and determine their historical periods more accurately.

References

1. Hutt, A.: Introducing Persian Architecture. By Arthur Upham Pope. (Library of Introductions to Persian Art.), p. 115. Oxford University Press, London (1969). £ 1.25. J. Roy. Asiatic Soc. **105**(2), 170-1 (1973)
2. Bier, C.: The decagonal tomb tower at Maragha and its architectural context: lines of mathematical thought. Nexus Netw. J. **14**(2), 251–273 (2012). https://doi.org/10.1007/s00004-012-0108-6
3. Ardalan, N., Bakhtiar, L.: The Sense of Unity: The Sufi Tradition in Persian Architecture. University of Chicago Press (1973)
4. Helaine, S.: Contested cultural heritage: a selective historiography. In: Silverman, H. (ed.) Contested Cultural Heritage, pp. 1–49. Springer, New York (2010). https://doi.org/10.1007/978-1-4419-7305-4_1
5. Grigor, T.: "They have not changed in 2,500 years": art, archaeology, and modernity in Iran. In: Unmasking Ideology in Imperial and Colonial Archaeology: Vocabulary, Symbols, and Legacy. Ideas, Debates, and Perspectives, pp. 121–146 (2018)
6. Kou, F., Du, J., He, Y., Ye, L.: Social network search based on semantic analysis and learning. CAAI Trans. Intell. Technol. **1**(4), 293–302 (2016)
7. Garcia-Garcia, A., Orts-Escolano, S., Oprea, S., Villena-Martinez, V., Garcia-Rodriguez, J.: A review on deep learning techniques applied to semantic segmentation. arXiv preprint arXiv: 170406857 (2017)
8. Li, L.-J., Hao, Su., Lim, Y., Fei-Fei, Li.: Objects as attributes for scene classification. In: Kutulakos, K.N. (ed.) ECCV 2010. LNCS, vol. 6553, pp. 57–69. Springer, Heidelberg (2012). https://doi.org/10.1007/978-3-642-35749-7_5
9. Srinivas, S., Sarvadevabhatla, R.K., Mopuri, K.R., Prabhu, N., Kruthiventi, S.S., Babu, R.V.: A taxonomy of deep convolutional neural nets for computer vision. Front. Robot. AI **2**, 36 (2016)
10. Sadeghi, Z., Alizadehsani, R., Cifci, M.A., Kausar, S., Rehman, R., Mahanta, P., et al.: A brief review of explainable artificial intelligence in healthcare. arXiv preprint arXiv:230401543 (2023)
11. Khozeimeh, F., Alizadehsani, R., Shirani, M., Tartibi, M., Shoeibi, A., Alinejad-Rokny, H., et al.: ALEC: active learning with ensemble of classifiers for clinical diagnosis of coronary artery disease. Comput. Biol. Med. **158**, 106841 (2023)
12. Nasab, R.Z., Ghamsari, M.R.E., Argha, A., Macphillamy, C., Beheshti, A., Alizadehsani, R., et al.: Deep learning in spatially resolved transcriptomics: a comprehensive technical view. arXiv preprint arXiv:221004453 (2022)
13. Asgharnezhad, H., Shamsi, A., Alizadehsani, R., Khosravi, A., Nahavandi, S., Sani, Z.A., et al.: Objective evaluation of deep uncertainty predictions for COVID-19 detection. Sci. Rep. **12**(1), 1–11 (2022)
14. Llamas, J., Lerones, P.M., Medina, R., Zalama, E., Gómez-García-Bermejo, J.: Classification of architectural heritage images using deep learning techniques. Appl. Sci. **7**(10), 992 (2017)
15. Perez, H., Tah, J.H., Mosavi, A.: Deep learning for detecting building defects using convolutional neural networks. Sensors **19**(16), 3556 (2019)
16. Makantasis, K., Doulamis, N.D., Voulodimos, A.: Recognizing buildings through deep learning: a case study on half-timbered framed buildings in Calw city. In: VISIGRAPP (5: VISAPP) (2017)
17. Obeso, A.M., Benois-Pineau, J., Acosta, A.Á.R., Vázquez, M.S.G.: Architectural style classification of Mexican historical buildings using deep convolutional neural networks and sparse features. J. Electron. Imaging **26**(1), 011016 (2017)

18. Vailaya, A., Figueiredo, M.A., Jain, A.K., Zhang, H.-J.: Image classification for content-based indexing. IEEE Trans. Image Process. **10**(1), 117–130 (2001)
19. Schreiber, S., Agne, S., Wolf, I., Dengel, A., Ahmed, S.: DeepDeSRT: deep learning for detection and structure recognition of tables in document images. In: 2017 14th IAPR International Conference on Document Analysis and Recognition (ICDAR). IEEE (2017)
20. Zhang, G., Pan, Y., Zhang, L.: Deep learning for detecting building façade elements from images considering prior knowledge. Autom. Constr. **133**, 104016 (2022)
21. Nahavandi, D., Alizadehsani, R., Khosravi, A., Acharya, U.R.: Application of artificial intelligence in wearable devices: opportunities and challenges. Comput. Methods Programs Biomed. **213**, 106541 (2022)
22. Alizadehsani, R., Sharifrazi, D., Izadi, N.H., Joloudari, J.H., Shoeibi, A., Gorriz, J.M., et al.: Uncertainty-aware semi-supervised method using large unlabeled and limited labeled COVID-19 data. ACM Trans. Multimedia Comput. Commun. Appl. (TOMM) **17**(3s), 1–24 (2021)
23. Alizadehsani, R., Roshanzamir, M., Izadi, N.H., Gravina, R., Kabir, H.D., Nahavandi, D., et al.: Swarm intelligence in internet of medical things: a review. Sensors **23**(3), 1466 (2023)
24. Alizadehsani, R., Habibi, J., Bahadorian, B., Mashayekhi, H., Ghandeharioun, A., Boghrati, R., et al.: Diagnosis of coronary arteries stenosis using data mining. J. Med. Sig. Sens. **2**(3), 153 (2012)
25. Zangooei, M.H., Habibi, J., Alizadehsani, R.: Disease Diagnosis with a hybrid method SVR using NSGA-II. Neurocomputing **136**, 14–29 (2014)
26. Alizadehsani, R., Hosseini, M.J., Boghrati, R., Ghandeharioun, A., Khozeimeh, F., Sani, Z.A.: Exerting cost-sensitive and feature creation algorithms for coronary artery disease diagnosis. Int. J. Knowl. Discov. Bioinform. (IJKDB). **3**(1), 59–79 (2012)
27. Alizadehsani, R., Habibi, J., Sani, Z.A., Mashayekhi, H., Boghrati, R., Ghandeharioun, A., et al.: Diagnosis of coronary artery disease using data mining based on lab data and echo features. J. Med. Bioeng. **1**(1), 26–29 (2012)
28. Nematollahi, M.A., Askarinejad, A., Asadollahi, A., Salimi, M., Moghadami, M., Sasannia, S., et al.: Association and predictive capability of body composition and diabetes mellitus using artificial intelligence: a cohort study (2022)
29. Sharifrazi, D., Alizadehsani, R., Hoseini Izadi, N., Roshanzamir, M., Shoeibi, A., Khozeimeh, F., et al.: Hypertrophic cardiomyopathy diagnosis based on cardiovascular magnetic resonance using deep learning techniques. Colour Filtering (2021)
30. Shoeibi, A., Khodatars, M., Jafari, M., Ghassemi, N., Moridian, P., Alizadesani, R., et al.: Diagnosis of brain diseases in fusion of neuroimaging modalities using deep learning: a review. Inf. Fusion **93**, 85–117 (2022)
31. Sharifrazi, D., Alizadehsani, R., Joloudari, J.H., Shamshirband, S., Hussain, S., Sani, Z.A., et al.: CNN-KCL: automatic myocarditis diagnosis using convolutional neural network combined with k-means clustering (2020)
32. Joloudari, J.H., Hussain, S., Nematollahi, M.A., Bagheri, R., Fazl, F., Alizadehsani, R., et al.: BERT-deep CNN: state-of-the-art for sentiment analysis of COVID-19 tweets. arXiv preprint arXiv:221109733 (2022)
33. Moridian, P., Shoeibi, A., Khodatars, M., Jafari, M., Pachori, R.B., Khadem, A., et al.: Automatic diagnosis of sleep apnea from biomedical signals using artificial intelligence techniques: methods, challenges, and future works. Wiley Interdisc. Rev. Data Min. Knowl. Discov. **12**(6), e1478 (2022)
34. Alizadehsani, R., Roshanzamir, M., Abdar, M., Beykikhoshk, A., Zangooei, M.H., Khosravi, A., et al.: Model uncertainty quantification for diagnosis of each main coronary artery stenosis. Soft. Comput. **24**, 10149–10160 (2020). https://doi.org/10.1007/s00500-019-04531-0

35. Nahavandi, S., Alizadehsani, R., Nahavandi, D., Mohamed, S., Mohajer, N., Rokonuzzaman, M., et al.: A comprehensive review on autonomous navigation. arXiv preprint arXiv:221 212808 (2022)
36. Khalili, H., Rismani, M., Nematollahi, M.A., Masoudi, M.S., Asadollahi, A., Taheri, R., et al.: Prognosis prediction in traumatic brain injury patients using machine learning algorithms. Sci. Rep. **13**(1), 960 (2023)
37. Abedini, S.S., Akhavan, S., Heng, J., Alizadehsani, R., Dehzangi, I., Bauer, D.C., et al.: A critical review of the impact of candidate copy number variants on autism spectrum disorders. arXiv preprint arXiv:230203211 (2023)
38. Karami, M., Alizadehsani, R., Argha, A., Dehzangi, I., Alinejad-Rokny, H.: Revolutionizing genomics with reinforcement learning techniques. arXiv preprint arXiv:230213268 (2023)
39. Jafari, M., Shoeibi, A., Khodatars, M., Ghassemi, N., Moridian, P., Delfan, N., et al.: Automated diagnosis of cardiovascular diseases from cardiac magnetic resonance imaging using deep learning models: a review. arXiv preprint arXiv:221014909 (2022)
40. Roshanzamir, M., Alizadehsani, R., Roshanzamir, M., Shoeibi, A., Gorriz, J.M., Khosrave, A., et al.: What happens in face during a facial expression? Using data mining techniques to analyze facial expression motion vectors. arXiv preprint arXiv:210905457 (2021)
41. Ayoobi, N., Sharifrazi, D., Alizadehsani, R., Shoeibi, A., Gorriz, J.M., Moosaei, H., et al.: Time series forecasting of new cases and new deaths rate for COVID-19 using deep learning methods. Results Phys. **27**, 104495 (2021)
42. Sharifrazi, D., Alizadehsani, R., Roshanzamir, M., Joloudari, J.H., Shoeibi, A., Jafari, M., et al.: Fusion of convolution neural network, support vector machine and Sobel filter for accurate detection of COVID-19 patients using X-ray images. Biomed. Sig. Process. Control **68**, 102622 (2021)
43. Alizadehsani, R., Roshanzamir, M., Hussain, S., Khosravi, A., Koohestani, A., Zangooei, M.H., et al.: Handling of uncertainty in medical data using machine learning and probability theory techniques: a review of 30 years (1991–2020). Ann. Oper. Res. 1–42 (2021). https://doi.org/10.1007/s10479-021-04006-2
44. Bidhend, M.Q., Abdollahzadeh, M.M.: Architecture, Land, and Man Rereading, reviewing and criticism of Pirnia's suggested principles for Iranian architecture. Art Alchemy/Kimiaye-Honar **3**(12), 72–99 (2014)
45. Hassannataj Joloudari, J., Azizi, F., Nematollahi, M.A., Alizadehsani, R., Hassannatajjeloudari, E., Nodehi, I., et al.: GSVMA: a genetic support vector machine ANOVA method for CAD diagnosis. Front. Cardiovasc. Med. **8**, 2178 (2022)
46. Iqbal, M.S., Ahmad, W., Alizadehsani, R., Hussain, S., Rehman, R.: Breast cancer dataset, classification and detection using deep learning. In: Healthcare. MDPI (2022)
47. Alizadehsani, R., Roshanzamir, M., Abdar, M., Beykikhoshk, A., Khosravi, A., Nahavandi, S., et al.: Hybrid genetic-discretized algorithm to handle data uncertainty in diagnosing stenosis of coronary arteries. Expert. Syst. **39**(7), e12573 (2022)
48. Joloudari, J.H., Alizadehsani, R., Nodehi, I., Mojrian, S., Fazl, F., Shirkharkolaie, S.K., et al.: Resource allocation optimization using artificial intelligence methods in various computing paradigms: a review. arXiv preprint arXiv:220312315 (2022)
49. Kakhi, K., Alizadehsani, R., Kabir, H.D., Khosravi, A., Nahavandi, S., Acharya, U.R.: The internet of medical things and artificial intelligence: trends, challenges, and opportunities. Biocybern. Biomed. Eng. **42**(3), 749–771 (2022)
50. Alizadehsani, R., Eskandarian, R., Behjati, M., Zahmatkesh, M., Roshanzamir, M., Izadi, N.H., et al.: Factors associated with mortality in hospitalized cardiovascular disease patients infected with COVID-19. Immun. Inflamm. Dis. **10**(3), e561 (2022)
51. Mahami, H., Ghassemi, N., Darbandy, M.T., Shoeibi, A., Hussain, S., Nasirzadeh, F., et al.: Material recognition for automated progress monitoring using deep learning methods. arXiv preprint arXiv:200616344 (2020)

52. Shoushtarian, M., Alizadehsani, R., Khosravi, A., Acevedo, N., McKay, C.M., Nahavandi, S., et al.: Objective measurement of tinnitus using functional near-infrared spectroscopy and machine learning. PLoS ONE **15**(11), e0241695 (2020)
53. Javan, A.A.K., Jafari, M., Shoeibi, A., Zare, A., Khodatars, M., Ghassemi, N., et al.: Medical images encryption based on adaptive-robust multi-mode synchronization of chen hyper-chaotic systems. Sensors **21**(11), 3925 (2021)
54. Eskandarian, R., Alizadehsani, R., Behjati, M., Zahmatkesh, M., Sani, Z.A., Haddadi, A., et al.: Identification of clinical features associated with mortality in COVID-19 patients. Oper. Res. Forum (2023). https://doi.org/10.1007/s43069-022-00191-3
55. Joloudari, J.H., Mojrian, S., Nodehi, I., Mashmool, A., Zadegan, Z.K., Shirkharkolaie, S.K., et al.: Application of artificial intelligence techniques for automated detection of myocardial infarction: a review. Physiol. Measur. 1–16 (2022)
56. Islam, S.M.S., Talukder, A., Awal, M., Siddiqui, M., Umer, M., Ahamad, M., et al.: Machine learning approaches for predicting hypertension and its associated factors using population-level data from three south asian countries. Front. Cardiovasc. Med. **9**, 762 (2022)
57. Joloudari, J.H., Azizi, F., Nodehi, I., Nematollahi, M.A., Kamrannejhad, F., Mosavi, A., et al.: DNN-GFE: a deep neural network model combined with global feature extractor for COVID-19 diagnosis based on CT scan images. Easychair, Manchester (2021)
58. Khozeimeh, F., Sharifrazi, D., Izadi, N.H., Joloudari, J.H., Shoeibi, A., Alizadehsani, R., et al.: RF-CNN-F: random forest with convolutional neural network features for coronary artery disease diagnosis based on cardiac magnetic resonance. Sci. Rep. **12**(1), 11178 (2022)
59. Jafari, M., Shoeibi, A., Ghassemi, N., Heras, J., Khosravi, A., Ling, S.H., et al.: Automatic diagnosis of myocarditis disease in cardiac MRI modality using deep transformers and explainable artificial intelligence. arXiv preprint arXiv:221014611 (2022)
60. Moravvej, S.V., Alizadehsani, R., Khanam, S., Sobhaninia, Z., Shoeibi, A., Khozeimeh, F., et al.: RLMD-PA: a reinforcement learning-based myocarditis diagnosis combined with a population-based algorithm for pretraining weights. Contrast Media Mol. Imaging **2022**, 1–15 (2022)

A Quasi-extreme Reduction for Interval Transportation Problems

Elif Garajová[1,2(✉)] [ID] and Miroslav Rada[2,3] [ID]

[1] Faculty of Mathematics and Physics, Department of Applied Mathematics,
Charles University, Prague, Czech Republic
elif@kam.mff.cuni.cz
[2] Faculty of Informatics and Statistics, Department of Econometrics,
Prague University of Economics and Business, Prague, Czech Republic
[3] Faculty of Finance and Accounting, Department of Financial Accounting
and Auditing, Prague University of Economics and Business, Prague, Czech Republic
miroslav.rada@vse.cz

Abstract. Transportation problems provide a classic linear programming model used in many areas of operations research, such as inventory control, logistics or supply chain management. The goal of a transportation problem is to find a minimum-cost transportation plan for shipping a given commodity from a set of sources to a set of destinations. Since the input data of such models are not always known exactly in practice, we adopt the approach of interval programming, which handles uncertainty in the supply, demand and cost parameters by assuming that only lower and upper bounds on these quantities are given.

One of the main tasks in interval programming is to compute bounds on the values that are optimal for some realization of the interval coefficients. While the best optimal value of an interval transportation problem can be computed by a single linear program, finding the worst (finite) optimal value is a much more challenging task. For interval transportation problems that are immune against the "more-for-less" paradox, it was recently proved that the worst optimal value can be found by considering only quasi-extreme scenarios, in which all coefficients in the model but one are set to the lower or upper bounds. We strengthen the former result and show that an analogous property also holds true for general interval transportation problems. Then, we utilize the obtained characterization to derive an exact method for computing the worst optimal value.

Keywords: Transportation problem · Interval linear programming · Worst optimal value

1 Introduction

A transportation problem is a linear programming model encountered in many practical applications, in which the goal is to create a minimum-cost transportation plan for shipping a given commodity from a set of sources with a limited supply to a set of destination such that the required demands are satisfied. Various approaches to modeling and solving transportation problems affected by different sources of uncertainty have been proposed in the literature. We adopt the approach of interval programming [9], leading to the model of an interval transportation problem [3], in which the available supply, the customer demand and the transportation costs can be uncertain and independently perturbed within some given lower and upper bounds.

One of the essential problems in interval programming is computing the best and the worst value that is optimal for some choice of the interval data, also known as the optimal value range problem. In this paper, we focus on the task of computing the worst possible optimal value, which was recently proved to be NP-hard for a class of interval transportation problems [7]. Several heuristic algorithms for approximating the worst optimal value have been proposed in the literature (see [2] and references therein), as well as a nonlinear duality-based formulation for computing the value [8] and an exact decomposition method based on complementary slackness [5].

Stronger results were derived for a special class of interval transportation problems that are immune against the "more-for-less" transportation paradox, which occurs when the total transportation cost can be reduced by increasing the amount of goods to be shipped. For these immune problems, it was proved that the worst optimal value can be found by considering so-called quasi-extreme scenarios, in which all coefficients in the model but one are set to the lower or upper bounds [1]. Here, we show that an analogous property also holds true for general interval transportation problems without the immunity assumption. Furthermore, we utilize the generalized result to derive an exact method for computing the worst optimal value. This exact method decomposes the problem into exponential number of subproblems (see Theorem 6), that is much lower that the number of subproblems in the other decomposition method from paper [5].

2   Interval Transportation Problem

To represent the uncertain or inexact data, we use an interval cost matrix and interval vectors for the supply and demand. Using the interval programming model, we assume that lower and upper bounds on the data are known.

Given real matrices $\underline{C}, \overline{C} \in \mathbb{R}^{m \times n}$ satisfying $\underline{C} \leq \overline{C}$ (inequalities are understood entry-wise), we define an *interval matrix* as the set

$$\mathbf{C} = [\underline{C}, \overline{C}] = \{C \in \mathbb{R}^{m \times n} : \underline{C} \leq C \leq \overline{C}\}.$$

An *interval vector (box)* can be defined analogously.

Let us now formally introduce the model of an interval transportation problem. Consider the following data:

– a set of m *sources* (or supply centers) denoted by $I = \{1, \ldots, m\}$, such that each source $i \in I$ has a limited *supply* s_i,

- a set of n *destinations* (or customers) denoted by $J = \{1, \ldots, n\}$, such that each destination $j \in J$ has a *demand* d_j to be satisfied,
- a *unit transportation cost* c_{ij} of transporting one unit of goods from source i to destination j.

Further, assume that the supplies s_i, demands d_j and costs c_{ij} are uncertain quantities that can vary within the given non-negative intervals $\mathbf{s}_i = [\underline{s}_i, \overline{s}_i]$, $\mathbf{d}_j = [\underline{d}_j, \overline{d}_j]$ and $\mathbf{c}_{ij} = [\underline{c}_{ij}, \overline{c}_{ij}]$, respectively. The objective of the transportation problem is to find a minimum-cost transportation plan for shipping goods from the sources to the destinations while respecting the supply availability and the demand requirements.

From the given data, we can define the model of an *interval transportation problem* in the form

$$
\begin{aligned}
\text{minimize} \quad & \sum_{i \in I} \sum_{j \in J} [\underline{c}_{ij}, \overline{c}_{ij}] x_{ij} \\
\text{subject to} \quad & \sum_{j \in J} x_{ij} \leq [\underline{s}_i, \overline{s}_i], && \forall i \in I, \\
& \sum_{i \in I} x_{ij} = [\underline{d}_j, \overline{d}_j], && \forall j \in J, \\
& x_{ij} \geq 0, && \forall i \in I, j \in J,
\end{aligned}
\tag{ITP}
$$

as the set of all transportation problems in the same form with data within the given intervals. A particular transportation problem in the interval problem (ITP) is called a *scenario*.

A lot of attention in interval linear programming studies has been devoted to finding the best or the worst value that is optimal for some scenario of the interval program [9]. In the traditional sense, infinite values are also considered "optimal" for interval programs with unbounded or infeasible scenarios. However, in many applications it is more useful to only take into account the feasible scenarios, i.e. to compute the best and the worst *finite* optimal value [6].

For the (ITP) model, feasibility of a particular scenario depends only on whether the available supply is sufficient to satisfy the demand. Thus, the set of all feasible scenarios can be characterized as

$$
\mathcal{SD} = \left\{ (s, d) \in (\mathbf{s}, \mathbf{d}) : \sum_{i \in I} s_i \geq \sum_{j \in J} d_j \right\}.
\tag{1}
$$

Let $f(C, s, d)$ denote the optimal value of a scenario (a fixed transportation problem) with data (C, s, d). Then, we can define the *best optimal value* $\underline{f}$ (which is finite, since the problem is always bounded) and the *worst finite optimal value* $\overline{f}_{\text{fin}}$ of (ITP) as the values

$$
\underline{f}(\mathbf{C}, \mathbf{s}, \mathbf{d}) = \min \{ f(C, s, d) : (s, d) \in \mathcal{SD}, \ C \in \mathbf{C} \},
\tag{2}
$$

$$
\overline{f}_{\text{fin}}(\mathbf{C}, \mathbf{s}, \mathbf{d}) = \max \{ f(C, s, d) : (s, d) \in \mathcal{SD}, \ C \in \mathbf{C} \}.
\tag{3}
$$

It can be shown that the best optimal value $\underline{f}$ can be computed by solving a single linear program (see [8] for details) in the form

$$\text{minimize} \quad \sum_{i \in I} \sum_{j \in J} \underline{c}_{ij} x_{ij}$$

$$\text{subject to} \quad \sum_{j \in J} x_{ij} \leq \overline{s}_i, \qquad \forall i \in I,$$

$$\underline{d}_j \leq \sum_{i \in I} x_{ij} \leq \overline{d}_j, \qquad \forall j \in J, \tag{4}$$

$$x_{ij} \geq 0, \qquad \forall i \in I, j \in J.$$

On the other hand, computing the worst finite optimal value $\overline{f}_{\text{fin}}$ is more challenging.

3   Quasi-extreme Scenarios and Their Properties

We will now derive a characterization of the worst finite optimal value $\overline{f}_{\text{fin}}$ of (ITP) by reducing the interval problem to a finite subset of feasible scenarios. Since the variables in our model are non-negative, we can observe that the worst (finite) optimal value is always attained for the upper costs $\overline{c}_{ij}$ (see e.g. [5]). Therefore, we can fix the cost matrix to $\overline{C}$ and consider only interval supplies $\mathbf{s}_i$ and interval demands $\mathbf{d}_j$.

To derive the characterization, we use a similar approach as in [1], where an analogous result was proved for transportation problems that are immune against the more-for-less paradox. Here, we prove a more general result that also holds true for transportation problems possibly affected by the paradox.

Using the terminology of the aforementioned paper, we say that a given scenario $(s, d) \in (\mathbf{s}, \mathbf{d})$ is *supply-quasi-extreme*, if there exists at most one index $k \in I$ of a source such that

$$s_i \in \{\underline{s}_i, \overline{s}_i\} \text{ for each } i \in I \setminus \{k\}, \qquad \underline{s}_k \leq s_k \leq \overline{s}_k, \text{ and}$$

$$d_j \in \{\underline{d}_j, \overline{d}_j\} \text{ for each } j \in J.$$

Similarly, a scenario $(s, d) \in (\mathbf{s}, \mathbf{d})$ is called *demand-quasi-extreme*, if there exists at most one index $k \in J$ of a destination such that

$$d_j \in \{\underline{d}_j, \overline{d}_j\} \text{ for each } j \in J \setminus \{k\}, \qquad \underline{d}_k \leq d_k \leq \overline{d}_k, \text{ and}$$

$$s_i \in \{\underline{s}_i, \overline{s}_i\} \text{ for each } i \in I.$$

We refer to the set of all supply-quasi-extreme and demand-quasi-extreme scenarios collectively as *quasi-extreme scenarios* (note that we also allow completely extreme scenarios in the definitions).

The reduction characterizing the worst finite optimal value of (ITP) is based on a convexity result that was proved in [4] showing that the optimal value function is convex on the set of feasible scenarios. The result is restated in Theorem 1.

Theorem 1 (D'Ambrosio et al. [4]). *The function $f(\overline{C}, s, d)$ is convex on the set $\mathcal{SD}$.*

Using Theorem 1, we can now show that it is sufficient to only consider the set of quasi-extreme scenarios for computing the value $\overline{f}_{\text{fin}}$, instead of taking into account the entire set $\mathcal{SD}$.

Theorem 2. *There exists a quasi-extreme scenario for which the value $\overline{f}_{\text{fin}}$ is attained.*

Proof. The value $\overline{f}_{\text{fin}}$ can be computed as maximum of the function $f(\overline{C}, s, d)$ over the set of feasible scenarios $\mathcal{SD}$. By Theorem 1, the function $f(\overline{C}, s, d)$ is convex on the set $\mathcal{SD}$. The set $\mathcal{SD}$ is a polyhedron. Since the maximum of a convex function over a polyhedron can be found at one of the vertices, the worst optimal value $\overline{f}_{\text{fin}}$ is attained at a scenario corresponding to a vertex of $\mathcal{SD}$.

Furthermore, a vertex of the set $\mathcal{SD}$ is either a vertex of the interval box $(\mathbf{s}, \mathbf{d})$ or it lies on an edge of the interval box intersected by the hyperplane

$$\sum_{i \in I} s_i \geq \sum_{j \in J} d_j.$$

In the former case, the vertex corresponds to a completely extreme (and thus also quasi-extreme) scenario, in which $s_i \in \{\underline{s}_i, \overline{s}_i\}$ and $d_j \in \{\underline{d}_j, \overline{d}_j\}$ holds true for all $i \in I$, $j \in J$. In the latter case, the vertex corresponds to a quasi-extreme scenario, since it belongs to a 1-dimensional face of the interval box $(\mathbf{s}, \mathbf{d})$. $\quad\square$

Although Theorem 2 does not directly yield a finite set of scenarios, we can note that for a particular quasi-extreme scenario, the value of s_k or d_k needed for computing $\overline{f}_{\text{fin}}$ can be exactly determined based on the remaining values of supply and demand. To show this, we use following results from [2] describing monotonicity of the optimal objective values with respect to changes in the supply and demand vectors.

Lemma 3 (Cerulli et al. [2]). *If (s^1, d) and (s^2, d) are two scenarios of (ITP) with a fixed cost matrix $C \in \mathbb{R}^{m \times n}$ such that $s_i^1 \leq s_i^2$ for each $i \in I$, then $f(C, s^1, d) \geq f(C, s^2, d)$.*

Theorem 4 (Cerulli et al. [2]). *If (s, d^1) and (s, d^2) are two scenarios of (ITP) with a fixed cost matrix $C \in \mathbb{R}^{m \times n}$ such that $d_j^2 \leq d_j^1$ for each $j \in J$, then $f(C, s, d^2) \leq f(C, s, d^1)$.*

Using the properties stated in Lemma 3 and Theorem 4, we can now show that the remaining coefficient in a considered supply-quasi-extreme or demand-quasi-extreme scenario can be fixed to a particular value. This leads to a description of the worst finite optimal value $\overline{f}_{\text{fin}}$ by a finite number of quasi-extreme scenarios, which can then be used to design a method for computing the value.

Theorem 5. *Let the worst optimal value $\overline{f}_{\mathrm{fin}}(\overline{C}, \mathbf{s}, \mathbf{d})$ be attained for a quasi-extreme scenario $(s, d) \in \mathcal{SD}$ with $\underline{s}_k < s_k < \overline{s}_k$ for some $k \in I$ or $\underline{d}_k < d_k < \overline{d}_k$ for some $k \in J$. Then, $\overline{f}_{\mathrm{fin}}$ is also attained for the scenario (s', d') where*

$$
s'_k = \max\left\{ \underline{s}_k, \; \sum_{j \in J} d_j - \sum_{i \in I \setminus \{k\}} s_i \right\}, \quad s'_i = s_i \text{ for } i \in I \setminus \{k\}, \quad d' = d
$$

for a supply-quasi-extreme scenario (s, d), or,

$$
d'_k = \min\left\{ \overline{d}_k, \; \sum_{i \in I} s_i - \sum_{j \in J \setminus \{k\}} d_j \right\}, \quad d'_j = d_j \text{ for } j \in J \setminus \{k\}, \quad s' = s
$$

for a demand-quasi-extreme scenario (s, d).

Proof. Let us assume that the worst finite optimal value $\overline{f}_{\mathrm{fin}}$ is attained for a supply-quasi-extreme scenario $(s, d) \in \mathcal{SD}$. The latter part of the theorem for a demand-quasi-extreme scenario can be proved analogously.

Consider the modified scenario (s', d') from the statement of the theorem, which differs from (s, d) only in the supply value s'_k. Since the original scenario (s, d) was feasible, we have

$$
\sum_{j \in J} d_j - \sum_{i \in I \setminus \{k\}} s_i \leq s_k \leq \overline{s}_k
$$

and thus $s'_k \leq \overline{s}_k$. We also have $s'_k \geq \underline{s}_k$ by definition, therefore $(s', d') \in (\mathbf{s}, \mathbf{d})$ is a valid scenario of the original interval transportation problem.

Further, we need to show that $f(\overline{C}, s', d') = \overline{f}_{\mathrm{fin}}$ holds true. Firstly, we can see that the modified scenario (s', d') is also feasible and therefore has a finite optimal value, since

$$
\sum_{i \in I} s'_i = s'_k + \sum_{i \in I \setminus \{k\}} s_i \geq \left(\sum_{j \in J} d_j - \sum_{i \in I \setminus \{k\}} s_i \right) + \sum_{i \in I \setminus \{k\}} s_i = \sum_{j \in J} d_j = \sum_{j \in J} d'_j.
$$

By Lemma 3, we know that decreasing the value of supply at source k from s_k to s'_k while keeping the remaining values of supply and demand the same can only lead to an increase in the optimal objective value, i.e. we have

$$
f(\overline{C}, s, d) \leq f(\overline{C}, s', d').
$$

However, we assumed that $f(\overline{C}, s, d)$ is the worst finite optimal value $\overline{f}_{\mathrm{fin}}$, therefore we also have $f(\overline{C}, s', d') = \overline{f}_{\mathrm{fin}}$.   □

4   Quasi-extreme Reduction

By fixing the last free value in a quasi-extreme scenario from Theorem 2 to the value given by Theorem 5 we get a reduction for computing $\overline{f}_{\mathrm{fin}}$ by considering only a finite (albeit still quite large) number of quasi-extreme scenarios, instead of the possibly infinite set of all feasible scenarios $\mathcal{SD}$, which was used in the original definition (3).

Based on the obtained results, we can now formulate a finite method for finding $\overline{f}_{\mathrm{fin}}$ as follows:

1. For each $k \in I$, consider all supply-quasi-extreme scenarios $(\overline{C}, s, d)$ with $s_i \in \{\underline{s}_i, \overline{s}_i\}$ for $i \in I \backslash \{k\}$ and $d_j \in \{\underline{d}_j, \overline{d}_j\}$ for $j \in J$.
   (a) If for a given scenario (s, d) we have

$$\sum_{j \in J} d_j - \sum_{i \in I \backslash \{k\}} s_i > \overline{s}_k,$$

   then the scenario is infeasible and can be ignored.
   (b) Otherwise, set s_k as in Theorem 5 and compute the optimal value.
2. For each $k \in J$, consider all demand-quasi-extreme scenarios $(\overline{C}, s, d)$ with $d_j \in \{\underline{d}_j, \overline{d}_j\}$ for $j \in J \backslash \{k\}$ and $s_i \in \{\underline{s}_i, \overline{s}_i\}$ for $i \in I$.
   (a) If for a given scenario (s, d) we have

$$\sum_{i \in I} s_i - \sum_{j \in J \backslash \{k\}} d_j < \underline{d}_k,$$

   then the scenario is infeasible and can be ignored.
   (b) Otherwise, set d_k as in Theorem 5 and compute the optimal value.
3. The worst finite optimal value $\overline{f}_{\mathrm{fin}}$ is obtained as the maximum over all optimal values computed in steps 1b and 2b.

Theorem 6. *It is sufficient to examine at most $(m + n)2^{m+n-1}$ quasi-extreme scenarios to compute $\overline{f}_{\mathrm{fin}}$.*

Proof. For each of $m + n$ supplies and demands, one can set each of $m + n - 1$ other supplies or demands to its lower or upper bound. □

The following example illustrates the three different optimal values that we have considered for an interval transportation problem with their corresponding scenarios: the best optimal value $\underline{f}$, the traditional worst optimal value $\overline{f}$ and the worst finite optimal value $\overline{f}_{\mathrm{fin}}$, which was mainly discussed in this paper.

Example. Consider an interval transportation problem in the form (ITP) with a fixed cost matrix given by the data (taken from [10])

$$C = \begin{pmatrix} 63 & 54 & 16 & 93 & 65 \\ 51 & 39 & 11 & 78 & 46 \\ 75 & 60 & 83 & 2 & 21 \end{pmatrix},$$

$$\mathbf{s} = ([60, 120], [75, 150], [100, 200])^T, \tag{5}$$

$$\mathbf{d} = ([45, 90], [30, 60], [60, 120], [25, 50], [75, 150])^T.$$

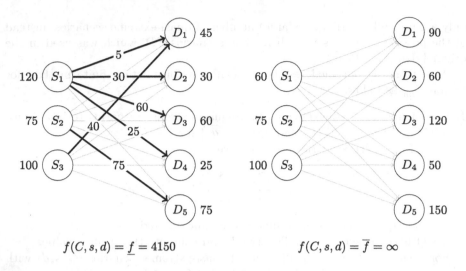

$$f(C, s, d) = \underline{f} = 4150 \qquad\qquad f(C, s, d) = \overline{f} = \infty$$

Fig. 1. The best and the worst (infinite) optimal value of the interval transportation problem with data (5) and the corresponding scenarios.

A scenario corresponding to the best optimal value $\underline{f} = 4150$, which can be computed by solving linear program (4), is depicted in Fig. 1. The interval problem also has an infeasible scenario, so the traditionally defined worst optimal value is $\overline{f} = \infty$ (see also Fig. 1).

A supply-quasi-extreme scenario corresponding to the worst finite optimal value $\overline{f}_{\text{fin}} = 11010$ is shown in Fig. 2. Note that the supply center S_3 provides a supply of 185, which is strictly inside the interval $\mathbf{s}_3 = [100, 200]$, while all other supplies and demands are set to the lower or upper bounds.

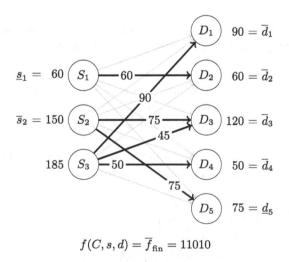

$$f(C, s, d) = \overline{f}_{\text{fin}} = 11010$$

Fig. 2. The worst finite optimal value of the interval transportation problem with data (5) and the corresponding supply-quasi-extreme scenario.

5   Conclusion

We addressed the computationally difficult problem of finding the worst finite optimal value of an interval transportation problem. We derived a reduction of the interval problem to a finite problem of computing the optimal values over a set of quasi-extreme scenarios, in which almost all supplies and demands (except for one) are set to their respective lower or upper bounds.

An improved reduction was formerly shown for interval transportation problems that are immune against the transportation paradox. Even though the problem of computing the worst optimal value is NP-hard, in general, it remains to be seen whether a more efficient reduction requiring a lower number of scenarios is possible, at least for some classes or forms of interval transportation problems.

Acknowledgements. The authors were supported by the Czech Science Foundation under Grant 23-07270S. The work of the first author was also supported by the grant SVV-2023-260699.

References

1. Carrabs, F., Cerulli, R., D'Ambrosio, C., Della Croce, F., Gentili, M.: An improved heuristic approach for the interval immune transportation problem. Omega **104**, 102492 (2021). https://doi.org/10.1016/j.omega.2021.102492
2. Cerulli, R., D'Ambrosio, C., Gentili, M.: Best and worst values of the optimal cost of the interval transportation problem. In: Sforza, A., Sterle, C. (eds.) ODS 2017. SPMS, vol. 217, pp. 367–374. Springer, Cham (2017). https://doi.org/10.1007/978-3-319-67308-0_37
3. Chanas, S., Delgado, M., Verdegay, J.L., Vila, M.A.: Interval and fuzzy extensions of classical transportation problems. Transp. Plan. Technol. **17**(2), 203–218 (1993). https://doi.org/10.1080/03081069308717511
4. D'Ambrosio, C., Gentili, M., Cerulli, R.: The optimal value range problem for the Interval (immune) Transportation Problem. Omega **95**, 102059 (2020). https://doi.org/10.1016/j.omega.2019.04.002
5. Garajová, E., Rada, M.: Interval transportation problem: feasibility, optimality and the worst optimal value. Cent. Eur. J. Oper. Res. 1–22 (2023). https://doi.org/10.1007/s10100-023-00841-9
6. Hladík, M.: The worst case finite optimal value in interval linear programming. Croat. Oper. Res. Rev. **9**(2), 245–254 (2018). https://doi.org/10.17535/crorr.2018.0019
7. Hoppmann-Baum, K.: On the complexity of computing maximum and minimum min-cost-flows. Networks 1–13 (2021). https://doi.org/10.1002/net.22060
8. Liu, S.T.: The total cost bounds of the transportation problem with varying demand and supply. Omega **31**(4), 247–251 (2003). https://doi.org/10.1016/S0305-0483(03)00054-9

9. Rohn, J.: Interval linear programming. In: Fiedler, M., Nedoma, J., Ramík, J., Rohn, J., Zimmermann, K. (eds.) Linear Optimization Problems with Inexact Data, pp. 79–100. Springer, Boston (2006). https://doi.org/10.1007/0-387-32698-7_3

10. Xie, F., Butt, M.M., Li, Z., Zhu, L.: An upper bound on the minimal total cost of the transportation problem with varying demands and supplies. Omega **68**, 105–118 (2017). https://doi.org/10.1016/j.omega.2016.06.007

Assignment of Unexpected Tasks in Embedded System Design Process Using Genetic Programming

Adam Górski[✉] and Maciej Ogorzałek

Department of Information Technologies, Jagiellonian University, Prof. Stanisława Łojasiewicza 11, Cracow, Poland
{a.gorski,maciej.ogorzalek}@uj.edu.pl

Abstract. Embedded systems need to execute some tasks. The designer needed to predict every behaviour of the system. However the problem appears when system meets unexpected situation. In some cases the system architecture cannot be modified or such operation is too expensive. In this paper we present a novel developmental genetic programming based method for assignment of unexpected tasks in embedded system design process. Our approach does not modify the system architecture. The proposed method evolves decision trees. The new individuals are obtained during evolution process after using genetic operators: mutation, crossover, cloning and selection. After genotype to phenotype mapping the ready system is obtained.

Keywords: Embedded systems · unexpected tasks · developmental genetic programming · optimization

1 Introduction

Nowadays we are surrounded by equipment which use embedded systems [1]. Modern cars [2], smart city solutions, traffic sign detection [3], cameras [4], IoT, agriculture solutions [5], home appliances such as washing machines or coffee machines and many others, all use embedded systems. Embedded systems are computer systems mostly microprocessor or microcontroller based dedicated to execute special types of tasks. They impact almost every aspect of daily live. Therefore it is very important to decrease the cost of embedded systems as much as possible. As a result the systems can be more affordable and further improve everyday life.

According to De Michelli and Gupta [6] Embedded system design process can be divided into three phases: modelling, validation and verification. One of most important processes in embedded system design is hardware/software co-synthesis [7]. During co-synthesis the architecture of the embedded system, and further task assignment and task scheduling is established. Co-synthesis algorithms can be divided into two types: constructive [8] and iterative improvement [9]. Constructive methods build the system by making separate decisions for every part of it. These algorithms have low complexity but can stop in local minima during optimization. Algorithms which allow to move back and change previous decisions are much more complex. Iterative improvements methods start from suboptimal solutions. Such algorithms search for an optimal system by making local decisions like allocation and deallocation of resources or reassignment of the tasks between resources. The methodologies can escape local minima on optimization landscape but the final results are still suboptimal. Evolution solutions like genetic algorithms [10] can also escape from local minima. However obtained results are very sensitive on parameters change. Very good results can be obtained using genetic programming [11, 12]. In such algorithms the way for obtaining the architecture was evolved instead of direct evolution of the architecture.

Genetic programming [13] is an extension of genetic algorithms [14]. It was applied to design electronic circuits [15], cosynthesis of embedded systems [16], optimization of residential energy flows [17], estimation of COVID-19 epidemic curves [18], and to solve many other problems. The main difference between genetic programming and genetic algorithm is that the genotype in genetic programming is a tree. In the nodes of the tree there are functions [19]. There are several types of genetic programming solutions. In linear genetic programming [20] the genotype tree is represented in a linear form. In cartesian genetic programming [21] the genotype is represented as a graph. Developmental genetic programming builds a genotype starting from an embryo. The embryo is an initial part of a solution.

The real problem in embedded system design appears when built system needs to execute some unexpected tasks [22]. In some of such situations the architecture cannot be modified or modification could be too expensive. Górski and Ogorzałek proposed a method to assign unexpected tasks without modification of the architecture [23]. However proposed algorithm was deterministic method and the results were suboptimal. In [22] it was proposed a deterministic algorithms for assignment of unexpected tasks for a group of embedded systems working together. The main disadvantages of the algorithm was that all of unexpected tasks that were investigated could appear only after predicted tasks were executed. In this paper we propose a genetic programming based algorithm that can assign unexpected tasks that can appear at any moment during the operation of the system. Due to the best of our knowledge this is the first implementation of genetic programming for solving the problem of unexpected tasks assignment. The paper is organized as follows: Sect. 2 presents basic assumptions, description of the embedded system architecture, its representation and definition of the problem of unexpected tasks. The Sect. 3 presents the algorithm. The last two sections describe experimental results and conclusions.

2 Assumptions

2.1 Embedded System

Embedded system is a computer system dedicated to execute some special tasks. Many modern embedded systems are implemented as distributed systems [24]. In such a situation the architecture contains many Processing Elements (PEs). There are two basic kinds of PEs: Programmable Processors (PPs) and Hardware Cores (HCs). PPs are universal resources. They are able to execute more than one task. Therefore the cost of such resources is relatively low. However the time of execution of the tasks by PPs is quite high. HCs are specialized resources able to execute only one task. They are very fast but usage of such resources very increases the overall cost of the system. The communication between PEs is provided by Communication Links (CLs).

Behaviour of the embedded system is described by an acyclic graph G = (V, E) called the task graph. Each node in this graph (v_i) represents a task. Each edge (e_{ij}) gives information about the data that needs to be transferred between two connected tasks (v_i and v_j). Transmission time (t_{ij}) between the tasks is given by formula (1):

$$t_{ij} = \frac{d_{ij}}{b_{CL}} \tag{1}$$

b_{CL} is a bandwidth and defines the amount of data that can be transferred through the communication link in a time unit, d_{ij} is the amount of data transferred between considered tasks.

2.2 Assignment of Unexpected Tasks

Unexpected tasks can appear at any moment of the operation of the embedded system. It means that the system has been designed and implemented and its architecture cannot be modified. In particular it is not possible to add any additional PP or CL. In such a situation the only possibility is to assign unexpected tasks to PPs which were chosen during cosynthesis process. All of those tasks must be inserted into the task graph. Figure 1 below contains an example of task graph with inserted 3 unexpected tasks: UT4, UT6 and UT8.

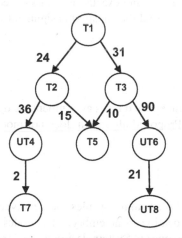

Fig. 1. Example of a task graph with inserted unexpected tasks.

The system described by the task graph above must execute 8 tasks. Tasks on the same level but on another paths can be executed in parallel. We assume that the cost and time of execution of each unexpected task is given. In Table 1 below we present an example of database for the graph of Fig. 1.

Table 1. Example of database.

Task	PP1 (x1)		PP2 (x3)		PP3 (x1)	
	c	t	c	t	c	t
T1	10	22	11	20	14	18
T2	5	15	6	13	8	10
T3	8	23	7	28	5	35
UT4	4	36	2	45	1	60
T5	12	18	10	25	13	17
UT6	6	40	8	35	10	31
T7	9	33	7	40	6	46
UT8	13	59	15	50	18	45

In this table there are also informations about number of each PP type in the system. In the example above the system has one PP1, three PP2s and one PP3. The overall cost of the final system (C_o) is described by (2):

$$C_o = C_i + \sum_{l=1}^{n} C_l + \sum_{k=1}^{m} C_k \tag{2}$$

where: n is the number of predicted tasks and m is the number of unexpected tasks. C_i is the initial cost of the architecture and execution of tasks executed by HCs. The goal in the investigated problem is to find the cheapest assignment which does not exceed the time constrains.

3   The Algorithm

To search for the optimal unexpected tasks assignment we decided to use developmental genetic programming. Proposed algorithm does not modify the architecture of the embedded system.

3.1   The Genotype

In accordance with genetic programming rules the genotype is a tree. The initial system (hardware and task allocation) is an embryo. The nodes of the tree contain system designing decisions or none option. System designing options are established for unexpected tasks and tasks executed by PPs. Tasks executed on HCs have no options. In the system should not exist any HC which does not execute a task. The system construction options are given in the Table 2 below. Each option has a probability of being chosen.

As it can be observed, the lowest probability equal to 0,1 have greedy options: the cheapest and the fastest assignment. The other options have probability equal to 0,2. To be sure that every task has been executed the number of nodes in the genotype is equal

Table 2. System construction options.

Number of option	Option	Probability
1	The cheapest assignment	0,1
2	The fastest assignment	0,1
3	Min (time * cost) for investigated task	0,2
4	The same as predecessor	0,2
5	Idle for the longest time	0,2
6	Lowest utilization	0,2

to the number of nodes in the task graph with inserted unexpected tasks. The structure of genotype (based on the task graph), for the same reason, needs to be the same for every individual and cannot be changed during the evolution process. Figure 2 presents an example of the genotype for the system described by the task graph from Fig. 1.

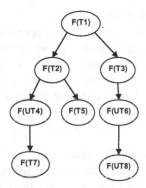

Fig. 2. Example of a genotype.

The genotype is a spanning tree of the task graph.

3.2 The Embryo

The embryo is an initial architecture of the embedded system with assignment of predicted tasks. The architecture is characterized by initial values: the time of execution of all the tasks and the cost of the system.

3.3 Evolution

At the beginning the initial population of genotypes is created. Number of individuals generated in every population is equal to Π and described by (3):

$$\Pi = \alpha * (m + n) \tag{3}$$

where n is the number of predicted tasks and m is the number of unexpected tasks. α is given by the designer and is a parameter which controls the size of the population.

Next generations are obtained using standard genetic operators: selection, mutation, crossover and cloning. After generating each population the results are ranked by cost. Selection chooses individuals with different probability described by (4):

$$P = \frac{\Pi - r}{\Pi} \tag{4}$$

The probability is depended on position r of the individual in the rank list. The best solutions have the biggest chance of being selected during evolution process. The probability of choosing the individual from the bottom of rank list is low, but not equal to 0.

Cloning copies ψ genotypes chosen by selection operator to new population. To avoid loss of the best individual, the best solution in current population is always cloned to new one. The number of individuals copied by cloning operator is described by (5):

$$\Psi = \delta * \Pi \tag{5}$$

where δ is a parameter given by the designer.

Mutation selects Ω genotypes with selection operator, then randomly one node is chosen. The operator substitutes the option in chosen the node to another one from table II. If the option in selected node is equal to none then another node is chosen. It is necessary to avoid a situation when in the system exist HCs that do not execute tasks. The number of solutions created using mutation operator is equal to (6):

$$\Omega = \beta * \Pi \tag{6}$$

where β is a parameter given by the designer.

Crossover operator selects φ individuals using selection. Then the individuals are randomly connected in pairs. Next the one crossing point is chosen. To be sure that every individual in new population will execute every task, the crossing point must be the same for both parent genotypes in each pair. The crossing point is a node. The last step is substitution of subtrees between every pair of genotypes. As the result two new solutions are created. The number of individuals generated by crossover is given by (7):

$$\Phi = \gamma * \Pi \tag{7}$$

where γ is a parameter given by the designer.

Parameters β, γ and δ control the evolution process. To have the same individuals in every generation the condition (8) must be satisfied:

$$\beta + \gamma + \delta = 1 \tag{8}$$

The algorithm stops if in next ε generations better individual was not found. ε is given be the designer.

4 Experimental Results

All the experiments were done on benchmarks with 10, 20, 30, 40 and 50 nodes. Values obtained using Developmental Genetic Programming (DGP) were compared with algorithm proposed by Górski and Ogorzałek [23]. For a problem of assignment of unexpected tasks in embedded system design process no genetic solution were investigated so far. In [22] authors investigated the problem for cooperation of group of embedded systems. Due to our best knowledge in this paper we propose the first genetic solution for such defined problem. The results of the comparison are presented in Table 3 below. T_{max} is the time constraint.

Table 3. Experimental results.

Graph	T_{max}	DGP 2023			Górski Ogorzałek 20216	
		c	t	Gen	c	t
10	1400	**954**	**1310**	**28**	1368	1369
20	3500	**1838**	**3477**	**52**	2370	3421
30	3500	**3231**	**3427**	**33**	3744	3115
40	4500	**4136**	**4463**	**52**	6148	1796
50	7000	**3211**	**6716**	**98**	6558	3548

The Graphical representation of obtained costs is presented in Fig. 3 below:

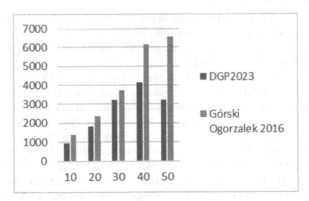

Fig. 3. Experimental results.

For smaller graphs (with 10 and 20 nodes) values obtained using DGP (either cost and time) are lower than values obtained using the second algorithm. For the benchmark with 30 nodes cost of the system obtained using DGP was better than cost obtained by the second algorithm however the value of time was higher. For bigger graphs (with 40 and 50 nodes) it can be observed that the costs in all of the systems obtained using DGP were much smaller than those which were obtained using the second algorithm. The difference for the graph with 50 nodes was greater than 50%. The time of the system obtained by DGP is much bigger but as it was mentioned before the goal was to find the cheapest solution. Therefore the results suggest that developmental genetic programming for the investigated problem is more effective than existing methods.

5 Conclusions

Experimental results indicate on strong advantage in terms of effectiveness of the presented algorithm. For every benchmark test the best results were generated using our approach.

In the future we plan to check other genes and to modify the algorithm to allow it to evolve genotypes with different numbers of nodes. We also plan to provide other types of algorithms, especially iterative improvement, based on genetic programming.

References

1. Guo, C., Ci, S., Yang, Y.: A servey of energy consumption measurement in embedded system. IEEE Access **9**, 60516–60530 (2021)
2. Laohapensaeng, T., Chaisricharoen, R., Boonyanant, P.: Evaluation system for car engine performance. In: Proceedings of Joint International Conference on Digital Arts, Media and Technology with ECTI Northern Section Conference on Electrical, Electronics, Computer and Telecommunication Engineering, Thailand, pp. 322–326 (2021)
3. Lopez-Montiel, M., Orozco-Rosas, U., Sanchez-Adame, M., Picos, K., Ross, O.H.M.: Evaluation method of deep learning-based embedded systems for traffic sign detection. IEEE Access **9**, 101217–101238 (2021)
4. Wolf, W., Ozer, B., Lv, T.: Smart cameras as embedded systems. Computer **35**, 48–53 (2002)
5. Saddik, A., Latif, R., El Ouardi, A., Elhoseny, M., Khelifi, A.: Computer development based embedded systems in precision agriculture: tools and application. Acta Agric. Scand. Sect. B Soil Plant Sci. **72**(1), 589–611 (2022)
6. De Micheli, G., Gupta, R.: Hardware/software co-design. Proc. IEEE **95**(3), 349–365 (1997)
7. Górski, A., Ogorzałek, M.: Genetic programming based algorithm for HW/SW cosynthesis of distributed embedded systems specified using conditional task graph. In: Proceedings of the 10th International Conference on Sensor Networks, pp. 239–243 (2022)
8. Górski, A., Ogorzałek, M.: Genetic programming based constructive algorithm with penalty function for hardware/software cosynthesis of embedded systems. In: Proceedings of the 16th International Conference on Software Technologies, ICSOFT, pp. 583–588 (2021)
9. Oh, H., Ha, S.: Hardware-software cosynthesis of multi-mode multi-task embedded systems with real-time constraints. In: Proceedings of the International Workshop on Hardware/Software Codesign, pp. 133–138 (2002)
10. Dick, R.P., Jha, N.K.: MOGAC: a multiobjective genetic algorithm for the co-synthesis of hardware-software embedded systems. IEEE Trans. Comput. Aided Des. Integr. Circ. Syst. **17**(10), 920–935 (1998)
11. Deniziak, S., Gorski, A.: Hardware/software co-synthesis of distributed embedded systems using genetic programming. In: Hornby, G.S., Sekanina, L., Haddow, P.C. (eds.) ICES 2008. LNCS, vol. 5216, pp. 83–93. Springer, Heidelberg (2008). https://doi.org/10.1007/978-3-540-85857-7_8

12. Górski, A., Ogorzałek, M.: Genetic programming based iterative improvement algorithm for HW/SW cosynthesis of distributted embedded systems. In: Proceedings of the 10th International Conference on Sensor Networks, pp. 120–125 (2021)
13. Langdon, W.B.: Genetic programming convergence. In: Genetic Programming and Evolvable Machines (2021)
14. Goldberg, D.E., Holland, J.H.: Genetic algorithms and machine learning. Mach. Learn. $3(2)$, 95–99 (1988)
15. Koza, J.R., Bennett III, F.H., Lohn, J., Dunlap, F., Keane, M.A., Andre, D.: Automated synthesis of computational circuits using genetic programming. In: Proceedings of the IEEE Conference on Evolutionary Computation (1997)
16. Górski, A., Ogorzałek, M.: Adaptive iterative improvement GP-based methodology for HW/SW co-synthesis of embedded systems. In: Proceedings of the 7th International Joint Conference on Pervasive and Embedded Computing and Communication Systems, Spain, pp. 56–59 (2017)
17. Kefer, K., et al.: Simulation-based optimization of residential energy flows using white box modeling by genetic programming. Energy Build. **258**, 111829 (2022)
18. Andelić, N., Baressi Šegota, S., Lorencin, I., Mrzljak, V., Car, Z.: Estimation of COVID-19 epidemic curves using genetic programming algorithm. Health Inform. J. $27(1)$ (2021)
19. Poli, R., Langdon, W., McPhee, N.: A field guide to genetic programming (2008). http://lulu. com. http://www.gp-field-guide.org.uk
20. Banzhaf, W., Nordin, P., Keller, R., Francone, F.: Genetic Programming - An Introduction. On the Automatic Evolution of Computer Programs and Its Application. dpunkt/Morgan Kaufmann, Heidelberg/San Francisco (1998)
21. Miller, J.F.: Cartesian genetic programming. In: Miller, J. (ed.) Cartesian Genetic Programming, pp. 17–34. Springer, Heidelberg (2011). https://doi.org/10.1007/978-3-642-173 10-3_2
22. Górski, A., Ogorzałek, M.: Assignment of unexpected tasks for a group of embedded systems. IFAC-PapersOnLine **51**(6), 102–106 (2018)
23. Górski, A., Ogorzałek, M.: Assignment of unexpected tasks in embedded system design process. Microprocess. Microsyst. **44**, 17–21 (2016)
24. Górski, A., Ogorzałek, M.: Adaptive GP-based algorithm for hardware/software co-design of distributed embedded systems. In: Proceedings of the 4th International Conference on Pervasive and Embedded Computing and Communication Systems, Portugal, pp. 125–130 (2014)

Improving Handwritten Cyrillic OCR by Font-Based Synthetic Text Generator

Ivan Gruber[(✉)] [iD], Lukáš Picek[iD], Miroslav Hlaváč[iD], Petr Neduchal[iD], and Marek Hrúz[iD]

Department of Cybernetics and New Technologies for the Information Society, Technická 8, 301 00 Pilsen, Czech Republic
grubiv@ntis.zcu.cz

Abstract. In this paper, we propose a straight-forward and effective Font-based Synthetic Text Generator (FbSTG) to alleviate the need for annotated data required for not just Cyrillic handwritten text recognition. Unlike standard GAN-based methods, the FbSTG does not have to be trained to learn new characters and styles; all it needs is the fonts, the text, and sampled page backgrounds. In order to show the benefits of the newly proposed method, we train and test two different OCR systems (Tesseract, and TrOCR) on the Handwritten Kazakh and Russian dataset (HKR) both with and without synthetic data. Besides, we evaluate both systems' performance on a private NKVD dataset containing historical documents from Ukraine with a high amount of out-of-vocabulary (OoV) words representing an extremely challenging task for current state-of-the-art methods. We decreased the CER and WER significantly by adding the synthetic data with the TrOCR-Base-384 model on both datasets. More precisely, we reduced the relative error in terms of CER/WER on (i) HKR-Test1 with OoV samples by around 20%/10%, and (ii) NKVD dataset by 24% CER and 8% WER. The FbSTG code is available at: https://github.com/mhlzcu/doc_gen.

Keywords: handwritten optical character recognition · Cyrillic · handwritten text generation · synthetic data · Tesseract · TrOCR · out-of-vocabulary

1 Introduction

Parts of human history are preserved as written documents. Before the advent of computers, many such documents were handwritten, typewritten or both.

This research was supported by the Ministry of Culture Czech Republic, project No. DG20P02OVV018. The work described herein has also been supported by the Ministry of Education, Youth and Sports of the Czech Republic, Project No. LM2023062 LINDAT/CLARIAH-CZ. Access to computing and storage facilities owned by parties and projects contributing to the National Grid Infrastructure MetaCentrum provided under the programme "Projects of Large Research, Development, and Innovations Infrastructures" (CESNET LM2015042), is greatly appreciated.

H. Moosaei et al. (Eds.): DIS 2023, LNCS 14321, pp. 102–115, 2024.
https://doi.org/10.1007/978-3-031-50320-7_8

Thanks to their digitization, we can not only ensure their preservation but also make them available to the general public. However, digitization itself is only the first step. The organization of digital documents is also important and can enable complex handling of the documents. To be able to organize the documents, multiple additional steps are necessary. One of the most important, but also one of the hardest steps is the Optical Character Recognition (OCR). Nevertheless, this step builds the foundation for the advanced tasks of document processing. When the documents are of high quality, modern OCR systems work very well [17,19,21]. Unfortunately, the domain of historical documents, and especially handwritten ones, is not covered on the same level, yet. The large variety of styles of writing imposes new challenges for the OCR systems. The standard approach of finetuning an OCR system to the specific type of documents can largely improve its performance. However, this step generally requires a large amount of annotated data. Unfortunately, data acquisition can be very time-consuming and expensive.

In this paper, we explore the possibility of synthetic data generation to mitigate the amount of annotated data needed for OCR system finetuning. To address this problem, we propose a novel Font-based Synthetic Text Generator and demonstrate that it lowers the character error rate (CER) and the word error rate (WER) on two datasets: the Handwritten Kazakh and Russian (HKR) dataset [24], and our private NKVD dataset. In the first step, we use two different OCR systems, which are first trained on the HKR dataset only, and evaluated on both HKR and NKVD datasets. In the second step, we train both OCR systems once again, but we provide them with additional synthetic data. Such trained OCRs are evaluated the same way as in step one. We observe consistent improvement in CER and WER while using additional synthetic data, especially on HKR-Test1 and NKVD datasets, which both contain OoV words.

The main contributions of this paper are as follows: (i) We propose Font-based Synthetic Text Generator, and (ii) we show that the addition of synthetic data into the train set consistently improves the results.

2 Related Work

In this section, we review literature related to handwritten text recognition, and synthetic handwritten text generation.

2.1 Handwritten Text Recognition

Most of the modern approaches for handwritten text recognition are based on deep neural networks. Graves *et al.* [11] first proposed the usage of Bidirectional Long Short-Term Memory (BiLSTM) with Connectionist Temporal Classification (CTC) loss. The combination of recurrent network with CTC loss then became the golden standard in handwritten recognition for many years [20,30,34]. Recently, the interest in attention based approaches have arisen. Works [3,18,31] proposed sequence-to-sequence models with attention. With the

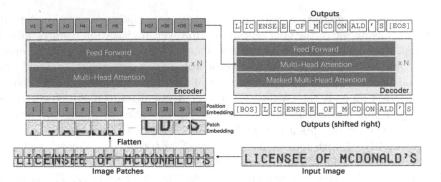

Fig. 1. TrOCR architecture flowchart overview. Illustration taken from the original paper [21].

emerge of the Transformers, methods based on them also became increasingly popular [17,33].

In this paper, we employ two methods based on Long Short-Term Memory (LSTM) and Transformer architecture – both described in more detail below.

Tesseract OCR [29] is an end-to-end OCR engine that uses—since version 4.0—LSTM-based OCR method and focuses on line detection as it is primarily intended to read typewritten documents. The Tesseract was initially developed by Hewlett-Packard and was open-sourced back in 2005. Since then, Google is the main contributor.

TrOCR [21] is a novel approach with state-of-the-art accuracy for both printed and handwritten text recognition. TrOCR is an encoder-decoder architecture where a pre-trained image transformer, e.g., ViT [9], DeIT [32] or BEiT [2], is proposed to be an encoder, and a standard NLP pre-trained model, e.g., RoBERTa [22] and XLM-RoBERTa [6], is used as a decoder (Fig. 1).

The input of an encoder is expected to be an image with 384 × 384 input resolution and a patch size of 16 × 16. The original paper introduced two different TrOCR architectures ("base" and "large") based on the number of layers and their width, and the number of heads. The base model's encoder has 12 layers with 768 hidden dimension and 12 heads, while the Large model has 24 layers with 1024 hidden dimension and 12 heads. We also test a Small model with 384 hidden dimension, depth of 12 and 6 heads. The decoder in the base and small models has 6 layers, 512 hidden dimension and 8 attention heads while in the large model it has 12 layers, 1024 hidden dimension and 16 heads. The framework itself does not restrict transformer architectures as feature-extractor nor as decoders. However, only the combination of the transformers provides state-of-the-art performance in terms of Character Error Rate (CER).

2.2 Synthetic Text Generation

Synthetic text generation algorithms can be divided into two main groups: font-based, and neural-network-based.

The font-based algorithms utilize font-rendering procedures to render text into an image, with possible additional augmentations. Typical representatives can be found in works [16], and [14], however, in contrast with our proposed method, these works are focused on text generation in the wild, not on the full-text documents.

The neural-network-based algorithms are nowadays usually based on LSTM [15] or on Generative Adversarial Networks (GANs) [7]. LSTM-based approaches [12] can generate fairly realistic-looking text, however, changing the font style of the generated text is not trivial. In recent years, LSTM-based approaches have been abandoned as the scientific community prefers approaches based on generative adversarial networks. Majority of modern approaches are based on the GANs. GAN-based can generate very realistic-looking handwritten text with the ability to easily change the font style. The quality of the generated texts is generally higher than the category of approaches based on recurrent neural networks. Currently, the most advanced methods at the time of writing this research are ScrabbleGAN [10], JokerGAN [36], and the method presented by Davis *et al.* [8].

3 Font-Based Synthetic Text Generator

Modern synthetic text generators are primarily based on GANs. Although these systems can generate good-looking text, they are conditioned by the character set used during the training. To generate novel characters, the system needs to be re-trained. A relatively high number of training samples is needed to achieve this, and many languages simply do not have the desired amount of labeled data available. We provide an example in Fig. 3.

Because we want the generator to produce Cyrillic characters, we opted for a font-based system. The other available font-based approaches are focused on texts in the wild, nevertheless, in this work, we are focusing on full-text generation to imitate real historical documents as much as possible. Although the font is deterministic in the appearance of individual characters, we make up for it

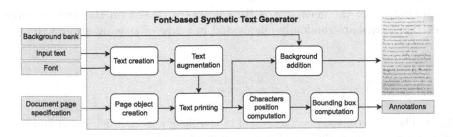

Fig. 2. Flowchart of proposed FbSTG framework for synthetic text generation.

by using augmentation techniques we describe in the following paragraphs. Furthermore, we do not need to train the system. We just need the textual corpora and proper fonts to generate labeled data.

Font-based Synthetic Text Generator (Fig. 2) is our software written in Python. Synthetic data are generated utilizing a pipeline that supports global and local augmentations, background injection, handwritten and typewritten fonts with varying sizes, and document layout generation.

The pipeline of the generator takes 4 inputs: (a) input text, (b) font, (c) background, and (d) configuration of the page. First, we create the page object based on the input configuration which specifies page size, page layout, size of boxes, and line spacing. Then, we create the input text using the font (TrueType and OpenType fonts are supported), apply augmentations and print this text on the page. Each pixel position in every character is recorded during printing. Afterward, we add a realistic background randomly chosen from the background bank and compute the bounding boxes of each character/word/line from the pixel positions. Output is the document with several lines of text and the annotations consisting of texts and bounding boxes of the corresponding text lines.

Augmentations were designed with the common distortions of documents in mind and are of several types. The positional augmentations change positions for each character; often used with typewritten fonts, but to a small degree, it works for handwritten fonts. The Perlin noise [25] is used to augment the texture in order to increase robustness by producing patterns with more realistic gradients between neighbouring pixels. A mask of the same size as the augmented character is generated and added to the character mask. The ratio of noise to the original mask is adjustable. Both types of augmentations can be generated on the level of font, individual characters, or a combination of both, allowing to simulate deviations in individual characters caused by typewriter or construction and deviations caused by the strength of individual strokes. The same applies to handwritten characters. We visualize the example augmentations in Fig. 4.

The process of generating individual characters is implemented with the preservation of the knowledge of the location of individual pixels in each character. This knowledge is then utilized to create bounding boxes for individual characters, words, and lines as can be seen in Fig. 5. The generator also checks the actual characters printed on the line and saves the information about the printed text. In case the text is longer than the size of the document only the printed characters are saved as annotations.

The output of the generator is then processed based on the line bounding boxes. The boxes are randomly augmented in terms of spacing around the text

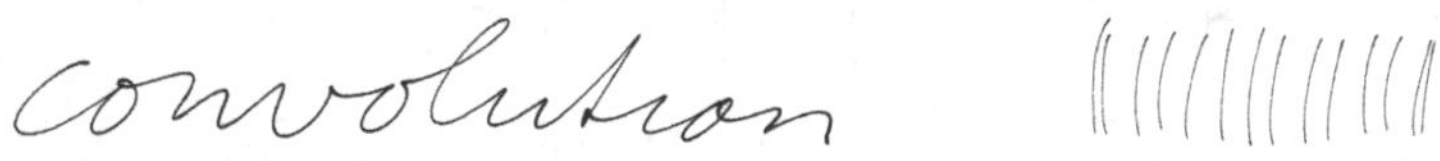

Fig. 3. The text "convolution" generated in Cyrillic ("коволюція", right) and Latin (left) by the system produces by GAN-based method [12].

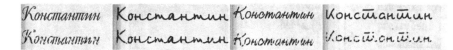

Fig. 4. Examples of automatically generated handwritten text with (bottom) and without (top) augmentations.

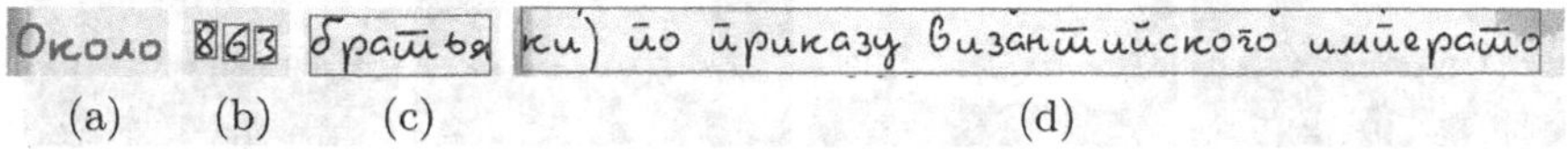

Около 863	братья	ки) по приказу византийского императо
(a) (b)	(c)	(d)

Fig. 5. Various annotations for text recognition. We differentiate annotations for (a) pixels, (b) letters (c) words and (d) lines.

and then extracted as separate images. The annotations are saved in a separated CSV file. Examples of synthetic data can be seen in Fig. 8.

4 Data

In this section we review in detail real-world Handwritten Kazakh and Russian (HKR) dataset and our NKVD dataset as well as data we generated using our synthetic document generator described in Sect. 3.

4.1 HKR – Handwritten Kazakh and Russian Database

The HKR database for text recognition is a collection of images with handwritten words in Cyrillic and their corresponding transcriptions into machine text. It was collected similarly to the IAM dataset [23] by using a form with pre-defined texts to be filled by writers. The authors report that the database consists of more than 1 500 filled forms written by 200 writers. There are approximately 63 000 sentences, more than 715 699 symbols and approximately 106 718 words. The character set consists of 33 symbols which are the same for both languages and 9 symbols that are exclusive for the Kazakh. The authors provide official splits of the dataset into Train (70%), Validation (15%) and Test (15%). The Test set is further split into two subsets – Test1 and Test2 – of equal cardinality. The Test1, is composed of out-of-vocabulary words – words not present in the train nor validation set. The Test2 contains words from the train set but written by writers excluded from the train set.

4.2 NKVD Dataset

The NKVD dataset contains historical documents found in the archives of NKVD (People's Commissariat for Internal Affairs) in the post-soviet states which are related to the history of Ukraine and (former) Czechoslovakia. The dataset was

Fig. 6. Example images from NKVD dataset.

constructed in a digitization effort led by the Czech Institute for Study of Totalitarian Regimes. The scanned documents often have poor image quality, posing a severe challenge for state-of-the-art OCR techniques. Furthermore, the scans regularly contain: (i) the text in multiple languages, (ii) various alphabets, e.g., Latin and Cyrillic, and (iii) typewritten and handwritten parts. In addition, there might be multiple documents in one scan, each with a different rotation, language, style, and quality (see Fig. 6).

The documents were semi-automatically pre-processed and manually annotated. The pre-processing and annotations consist of the following steps:

- **Main page detection and crop**: All scanned documents are feed-forwarded through the fine-tuned Faster R-CNN [27] to determine primary region.
- **Deskewing**: An automatic algorithm [13] based on the 2D Fourier transform is used to rotate a given scan to align the text with the x-axis of the image.
- **Document layout annotation**: Important parts of each document are annotated with a bounding box and appropriate label, e.g., typewritten heading or paragraph, handwritten heading or paragraph, image, and "other".
- **Text annotation**: All text blocks from the previous step are manually annotated, i.e., the text was rewritten by native speakers. In this paper, we experiment with handwritten heading and handwritten paragraphs only.
- **Line annotation**: Each text block is parsed into separate text lines. This step is necessary only because of the nature of used OCR algorithms.

In order to evaluate OCR performance for selected methods in OoV scenario, we prepared two test subsets – NKVD–1 and NKVD–2. The NKVD–1 contains 1295 text lines from the handwritten heading class containing titles, headings, page numbers, etc. The NKVD–2 includes 1525 text lines mainly from the handwritten paragraph class, containing longer multi-line texts.

4.3 Synthetic Data

We have generated two synthetic datasets using the generator described in Sect. 3. The size of the document page was 1560 × 2328. Then a line of text was generated with height given by an interval of (60–80) pixels. Our algorithm automatically calculates the corresponding font size to fit the line height. We have used four handwritten Cyrillic fonts. The fonts are randomly selected with uniform distribution to generate a single line of text on the page. The text is then generated with noise augmentations (random weight in the range of (0.3–0.7)) and small shift augmentations in the range of (0–3) pixel to preserve the connections between the individual characters. The line of text is then added to the document, and new lines are generated until the page is full. The page with text is blended with the chosen background through alpha channel blending. Background images are obtained using background extraction algorithm [5] from real scanned historical documents. Example backgrounds are visualized in Fig. 7.

The generated datasets differ in the distribution of characters and the length of sentences in source datasets. We have created a specific database of texts representing the same character distribution as in HKR and NKVD datasets by the process of distribution matching. We have randomly selected a starting sets of sentences from our internal dataset of Cyrillic texts, calculated the distribution of characters and then randomly select new candidates that can improve the match between the current distribution of the set and the desired distribution of target dataset. We calculate the relative distribution of characters. After the distributions are matched we randomly shuffle the set of sentences and create splits in the sizes of target datasets.

5   Experiments

In this section we first compare the performance of the chosen OCR systems with state-of-the-art results. Then we show the influence of synthetic data on the OCRs' accuracy while using different training protocols. Moreover, we evaluate these models on NKVD dataset.

5.1   Metrics

For the performance evaluation we use character error rate (CER) and word error rate (WER). We calculate CER calculated as Levenshtein distance:

Fig. 7. Examples of available background images—manually prepared samples.

$$\text{CER} = \frac{S_c + I_c + D_c}{N_c}, \tag{1}$$

where S_c is the number of substitutions, I_c is the number of insertions, D_c is the number of deletions, and N_c is the total number characters in the ground truth. Similarly, WER is calculated as:

$$\text{WER} = \frac{S_w + I_w + D_w}{N_w}, \tag{2}$$

where S_w is the number of substitutions, I_w insertions, D_w deletions, and N_w is the total number words in the ground truth.

5.2   Training Setup

Tesseract in version 4 and newer provides set of tools for training – primarily fine-tuning but it is possible to train from scratch as well. In our experiments we perform fine-tuning of Russian trained model with the following hyperparameters: (i) number of epochs = 30, (ii) initial learning rate = $5e^{-4}$, (iii) SGD optimizer with momentum = 0.5.

TrOCR To fine-tune the TrOCR on a custom dataset, we used the Transformers library introduced by HuggingFace [35] which includes various pre-trained models for handwritten text recognition. Hyperameters are shared for all the experiments featuring TrOCR and set as follows:

- Batch Size was set to 256 and accordingly accumulated based on the available number of GPUs and vRAM. Most of the time, we accumulated 16 gradients of a batch size of 16 from two GPUs.
- Learning Rate was set to $1e^{-5}$ and decreased linearly to $1e^{-9}$ for 50 epochs.
- In order to increase the training speed, we used FP16 even though there is a drawback in terms of slightly worst CER performance.

Remaining hyper-parameters used default settings of a *TrainingArguments* class.

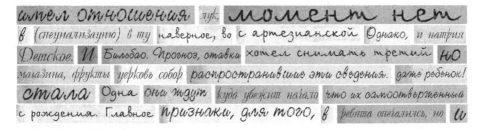

Fig. 8. Text generated by proposed font-based synthetic text generator (FbSTG).

5.3   State-of-the-Art Comparison

In this subsection, we compare the performance of the selected OCR approaches with other state-of-the-art methods, see Table 1. Both TrOCR and Tesseract were trained while using the HKR training set only. The goal of this experiment is not to reach new state-of-the-art, but rather to show the suitability of the selected OCRs for the given task. It can be seen that TrOCR reaches competitive results setting the new state-of-the-art for HKR–Test2. Tesseract provides noticeably worse results, especially in the WER metric. We believe it primarily stems from Tesseract's pipeline which is focused on typewritten texts.

5.4   Synthetic Data Effect

As mentioned earlier, to demonstrate the strength of the synthetic data generator, we trained TrOCR and Tesseract not only on the HKR training set, but we also enriched the original train set with synthetic data and trained the OCRs while using the original training protocols.

We test different ratios between the amount of the original and synthetic data. The results can be found in Table 2. It can be seen that the synthetic data consistently improves the results of TrOCR on the Test1 dataset, while it maintains similar performance on Test2. Test1 contains out-of-vocabulary words, which indicates that the synthetic data improves the generalization ability of the OCR system.

On the other hand, the synthetic data has minimal influence on the performance of Tesseract. The biggest results improvement is 0.70% of absolute error for Test1 and 0.74% of absolute error for Test2. In Table we report only the training protocol with the biggest improvement overall. We argue this is caused by the lower generalization capacity of LSTMs compared to Transformers. This causes an inability of Tesseract to fully utilize synthetic data.

All the trained OCR systems were also evaluated on the NKVD–2 dataset. Due to the challenging nature of the dataset, the overall accuracy is significantly

Table 1. Performance comparison for selected OCR methods on HKR test sets.

Architecture	HKR–Test1		HKR–Test2	
	CER [%]	WER [%]	CER [%]	WER [%]
1D-LSTM [26]	22.91	34.64	20.73	33.93
Gated-CRNN [4]	10.50	29.63	10.55	29.80
Attention-GatedCNN-BGRU [1]	**6.31**	**23.69**	4.13	18.91
StackMix-OCR [28]	×	×	3.49	13.00
Tesseract [19]	22.65	72.52	19.37	63.78
TrOCR–Small [21]	42.51	65.35	3.67	**7.35**
TrOCR–Base [21]	14.72	38.45	**2.80**	7.80
TrOCR–Large [21]	10.86	34.20	3.18	10.19

Table 2. Synthetic data experiment. Performance comparison for various real-to-synthetic ratios. Tested with TrOCR–Base and Tesseract.

Real:Synt ratio	HKR–Test1		HKR–Test2		NKVD–2	
	CER [%]	WER [%]	CER [%]	WER [%]	CER [%]	WER [%]
1:0	14.72	38.45	2.80	7.80	66.85	96.42
1:0.25	11.59	35.21	3.02	8.57	52.64	89.90
1:0.5	11.19	34.28	2.71	8.17	52.20	89.58
1:0.75	11.22	34.31	2.70	8.05	51.11	88.68
1:1	**10.88**	34.79	2.62	8.11	51.75	88.96
1:2	11.62	**34.29**	**2.58**	**7.77**	**51.08**	**88.43**
1:0 (Tesseract)	22.65	72.52	19.37	63.78	52.36	91.46
1:2 (Tesseract)	22.05	69.89	18.57	62.18	53.12	90.67

worse than the accuracy of the HKR dataset, see Table 2. Nevertheless, from the results, it can be observed the same phenomenon as for HKR–Test1 – the usage of the synthetic data significantly improves TrOCR results. This is expected behavior because the NKVD dataset also contains out-of-vocabulary words. The results of Tesseract are only slightly better, which, once again, we believe is caused unsuitability of Tesseract for handwritten texts.

In addition, we experiment with different distribution of synthetic data. We generate new dataset based on NKVD dataset distribution. All experiments are performed with TrOCR–Base trained on a training set with a 1:0.25 ratio only, see Table 3. Small improvement of results on the NKVD–2 dataset can be seen while using the synthetic data adapted directly on this dataset, however, it slightly worse results on HKR–Test1. We argue that the ability of the text generator to adjust data distribution to the original dataset can be essential for reaching good results. Even bigger improvements can be observed while using both synthetic datasets combined.

Table 3. Synthetic data combination performance. Evaluated on TrOCR–Base.

Synthetic Data		HKR–Test1		NKVD–1		NKVD–2	
HKR	NKVD	CER [%]	WER [%]	CER [%]	WER [%]	CER [%]	WER [%]
×	×	14.72	38.45	71.24	101.16	66.85	96.42
✓	×	11.59	35.21	62.26	97.66	52.64	89.90
×	✓	11.92	35.11	63.60	97.06	51.93	88.96
✓	✓	12.05	34.37	62.12	96.79	50.40	88.11

6   Conclusion

In this paper, we have proposed a solution for the problem of handwritten OCR. Namely, we have presented the NKVD dataset as a representative of real historical document scans and experimentally show the deterioration of performance of OCR when compared to laboratory-like data. Next, we have shown the positive effect of adding synthetic data to the training set on CER and WER. For this purpose, we have devised a font-based synthetic data generator. The main advantage of using a font-based generator as opposed to a GAN-based one is that there is no need for training it. We provide the codes for the generator at https://github.com/mhlzcu/doc_gen. We observe a relative reduction of around 25% CER for both the HKR test set with out-of-vocabulary words as well as the NKVD dataset. Next, we analyzed the OCR errors and found out that the system confuses the lower and the upper case versions of the characters, which does not impact the semantics of the text. When we omit the case sensitivity we observe a relative reduction of CER by 7%–30%.

References

1. Abdallah, A., Hamada, M., Nurseitov, D.: Attention-based fully gated CNN-BGRU for Russian handwritten text. J. Imaging **6**(12), 141 (2020)
2. Bao, H., Dong, L., Wei, F.: BEiT: BERT pre-training of image transformers. arXiv preprint arXiv:2106.08254 (2021)
3. Bluche, T., Louradour, J., Messina, R.O.: Scan, attend and read: end-to-end handwritten paragraph recognition with MDLSTM attention. CoRR abs/1604.03286 (2016). http://arxiv.org/abs/1604.03286
4. Bluche, T., Messina, R.: Gated convolutional recurrent neural networks for multilingual handwriting recognition. In: 2017 14th IAPR International Conference on Document Analysis and Recognition (ICDAR), vol. 1, pp. 646–651. IEEE (2017)
5. Bureš, L., Neduchal, P., Hlaváč, M., Hrúz, M.: Generation of synthetic images of full-text documents. In: Karpov, A., Jokisch, O., Potapova, R. (eds.) SPECOM 2018. LNCS (LNAI), vol. 11096, pp. 68–75. Springer, Cham (2018). https://doi.org/10.1007/978-3-319-99579-3_8
6. Conneau, A., et al.: Unsupervised cross-lingual representation learning at scale. arXiv preprint arXiv:1911.02116 (2019)
7. Creswell, A., White, T., Dumoulin, V., Arulkumaran, K., Sengupta, B., Bharath, A.A.: Generative adversarial networks: an overview. IEEE Sig. Process. Mag. **35**(1), 53–65 (2018)
8. Davis, B., Tensmeyer, C., Price, B., Wigington, C., Morse, B., Jain, R.: Text and style conditioned GAN for generation of offline handwriting lines. arXiv preprint arXiv:2009.00678 (2020)
9. Dosovitskiy, A., et al.: An image is worth 16 × 16 words: transformers for image recognition at scale. arXiv preprint arXiv:2010.11929 (2020)
10. Fogel, S., Averbuch-Elor, H., Cohen, S., Mazor, S., Litman, R.: ScrabbleGAN: semi-supervised varying length handwritten text generation. In: Proceedings of the IEEE/CVF Conference on Computer Vision and Pattern Recognition, pp. 4324–4333 (2020)

11. Graves, A., Liwicki, M., Fernandez, S., Bertolami, R., Bunke, H., Schmidhuber, J.: A novel connectionist system for unconstrained handwriting recognition. IEEE Trans. Pattern Anal. Mach. Intell. **31**(5), 855–868 (2009). https://doi.org/10.1109/tpami.2008.137
12. Graves, A.: Generating sequences with recurrent neural networks. arXiv preprint arXiv:1308.0850 (2013)
13. Gruber, I., et al.: OCR improvements for images of multi-page historical documents. In: Karpov, A., Potapova, R. (eds.) SPECOM 2021. LNCS (LNAI), vol. 12997, pp. 226–237. Springer, Cham (2021). https://doi.org/10.1007/978-3-030-87802-3_21
14. Gupta, A., Vedaldi, A., Zisserman, A.: Synthetic data for text localisation in natural images. In: Proceedings of the IEEE Conference on Computer Vision and Pattern Recognition, pp. 2315–2324 (2016)
15. Hochreiter, S., Schmidhuber, J.: Long short-term memory. Neural Comput. **9**(8), 1735–1780 (1997)
16. Jaderberg, M., Simonyan, K., Vedaldi, A., Zisserman, A.: Synthetic data and artificial neural networks for natural scene text recognition. arXiv preprint arXiv:1406.2227 (2014)
17. Kang, L., Riba, P., Rusiñol, M., Fornés, A., Villegas, M.: Pay attention to what you read: non-recurrent handwritten text-line recognition. arXiv preprint arXiv:2005.13044. CoRR abs/2005.13044 (2020). http://arxiv.org/abs/2005.13044
18. Kang, L., Toledo, J.I., Riba, P., Villegas, M., Fornés, A., Rusiñol, M.: Convolve, attend and spell: an attention-based sequence-to-sequence model for handwritten word recognition. In: Brox, T., Bruhn, A., Fritz, M. (eds.) GCPR 2018. LNCS, vol. 11269, pp. 459–472. Springer, Cham (2019). https://doi.org/10.1007/978-3-030-12939-2_32
19. Kay, A.: Tesseract: an open-source optical character recognition engine. Linux J. **2007**(159), 2 (2007)
20. Krishnan, P., Dutta, K., Jawahar, C.: Word spotting and recognition using deep embedding. In: 2018 13th IAPR International Workshop on Document Analysis Systems (DAS), pp. 1–6 (2018). https://doi.org/10.1109/DAS.2018.70
21. Li, M., et al.: TrOCR: transformer-based optical character recognition with pre-trained models. arXiv preprint arXiv:2109.10282 (2021)
22. Liu, Y., et al.: RoBERTa: a robustly optimized BERT pretraining approach. arXiv preprint arXiv:1907.11692 (2019)
23. Marti, U.V., Bunke, H.: The IAM-database: an English sentence database for offline handwriting recognition. Int. J. Doc. Anal. Recogn. **5**(1), 39–46 (2002). https://doi.org/10.1007/s100320200071
24. Nurseitov, D., Bostanbekov, K., Kurmankhojayev, D., Alimova, A., Abdallah, A., Tolegenov, R.: Handwritten Kazakh and Russian (HKR) database for text recognition. Multimed. Tools Appl. **80**, 33075–33097 (2021). https://doi.org/10.1007/s11042-021-11399-6
25. Perlin, K.: An image synthesizer. SIGGRAPH Comput. Graph. **19**(3), 287–296 (1985). https://doi.org/10.1145/325165.325247
26. Puigcerver, J.: Are multidimensional recurrent layers really necessary for handwritten text recognition? In: 2017 14th IAPR International Conference on Document Analysis and Recognition (ICDAR), vol. 1, pp. 67–72. IEEE (2017)
27. Ren, S., He, K., Girshick, R., Sun, J.: Faster R-CNN: towards real-time object detection with region proposal networks. In: Advances in Neural Information Processing Systems, vol. 28 (2015)

28. Shonenkov, A., Karachev, D., Novopoltsev, M., Potanin, M., Dimitrov, D.: Stack-Mix and blot augmentations for handwritten text recognition. arXiv preprint arXiv:2108.11667 (2021)
29. Smith, R.: An overview of the tesseract OCR engine. In: Ninth International Conference on Document Analysis and Recognition (ICDAR 2007), vol. 2, pp. 629–633. IEEE (2007)
30. Stuner, B., Chatelain, C., Paquet, T.: Cohort of LSTM and lexicon verification for handwriting recognition with gigantic lexicon. CoRR abs/1612.07528 (2016). http://arxiv.org/abs/1612.07528
31. Sueiras, J., Ruiz, V., Sanchez, A., Velez, J.F.: Offline continuous handwriting recognition using sequence to sequence neural networks. Neurocomputing **289**(1), 119–128 (2018). https://doi.org/10.1016/j.neucom.2018.02.008
32. Touvron, H., Cord, M., Douze, M., Massa, F., Sablayrolles, A., Jégou, H.: Training data-efficient image transformers & distillation through attention. In: International Conference on Machine Learning, pp. 10347–10357. PMLR (2021)
33. Vaswani, A., et al.: Attention is all you need. In: Proceedings of the 31st International Conference on Neural Information Processing Systems, NIPS 2017, pp. 6000–6010. Curran Associates Inc., Red Hook (2017)
34. Wigington, C., Stewart, S., Davis, B., Barrett, B., Price, B., Cohen, S.: Data augmentation for recognition of handwritten words and lines using a CNN-LSTM network. In: 2017 14th IAPR International Conference on Document Analysis and Recognition (ICDAR), vol. 01, pp. 639–645 (2017). https://doi.org/10.1109/ICDAR.2017.110
35. Wolf, T., et al.: Transformers: state-of-the-art natural language processing. In: Proceedings of the 2020 Conference on Empirical Methods in Natural Language Processing: System Demonstrations, pp. 38–45. Association for Computational Linguistics (2020). https://www.aclweb.org/anthology/2020.emnlp-demos.6
36. Zdenck, J., Nakayama, H.: JokerGAN: memory-efficient model for handwritten text generation with text line awareness. In: Proceedings of the 29th ACM International Conference on Multimedia, pp. 5655–5663 (2021)

A Critical Node-Centric Approach to Enhancing Network Security

Essia Hamouda$^{(\boxtimes)}$ [ID]

California State University San Bernardino, San Bernardino, USA
ehamouda@csusb.edu
https://www.csusb.edu/profile/essia.hamouda

Abstract. In the realm of network analysis, the identification of critical nodes takes center stage due to their pivotal role in maintaining network functionality. These nodes wield immense importance, as their potential failure has the capacity to disrupt connectivity and pose threats to network security. This paper introduces an innovative approach to assess the vulnerability of these critical nodes by assessing their significance within the network structure. Through rigorous numerical analysis, our methodology not only demonstrates its effectiveness but also offers valuable insights into network dynamics. To enhance network robustness and, consequently, enhance network security, we formulate the network as a non-linear optimization problem. Our overarching objective is to determine the optimal security level, quantified as a resource allocation cost, for these critical nodes, ultimately aligning with our network security and robustness objectives.

Keywords: node centrality · cybersecurity · performance analysis · optimization · convexity

1 Introduction

Networks have generated significant attention in complex system research, due to their universal applicability in modeling a diverse array of natural and societal systems [4, 41]. This stems from the fundamental concept that a complex system can be effectively represented as a graph denoted as $G = (V, E)$, where V signifies a collection of components linked through edges E, as illustrated in Fig. 1.

However, as these networks continue to expand, they bring along the challenge of an increasing proliferation of cybersecurity threats. These threats can emanate from two distinct sources: random failures and targeted destruction [54]. Research indicates that numerous real-world systems, including the Internet, exhibit considerable heterogeneity [27], resulting in significant variation in the roles of nodes within network structure and function. While such networks demonstrate resilience against random failures, they are notably vulnerable to deliberate attacks.

H. Moosaei et al. (Eds.): DIS 2023, LNCS 14321, pp. 116–130, 2024.
https://doi.org/10.1007/978-3-031-50320-7_9

These attacks often center around specific nodes within communication networks, commonly referred to as critical nodes. The compromise or malicious actions involving these nodes can exert a substantial influence on network performance and functionality.

Generally, attackers initiate their strategies by exploiting vulnerabilities within a network, eventually zeroing in on these critical nodes. For example, successfully targeting vital Internet service providers like AT&T or Sprint servers could result in widespread disruptions, affecting the operations of millions of companies' websites and online services. Conversely, critical nodes can also serve as the central focus of protective measures and proactive surveillance, aimed at enhancing network security. Unusual traffic patterns within a network can serve as indicators of potential server or crucial network element breaches, which could lead to unauthorized data transmissions or denial-of-service attacks.

To ensure the uninterrupted operation of a network, it is imperative to investigate network vulnerabilities, identify critical nodes, maintain vigilant monitoring, and implement appropriate security measures. These actions are essential for preventing or mitigating potential attacks, ultimately safeguarding the normal functioning of the network.

The identification of critical nodes has garnered considerable interest and found broad application across various domains. It offers valuable utility in enhancing the resilience and security of numerous applications. For instance, the development of network routing applications [25, 26] often revolves around the identification of a specific set of nodes for establishing a path between a source and a destination. Consequently, the security of data transmission is intricately linked to the criticality of the selected nodes within this path. The compromise of even a single node along the path, such as nodes u or v in Fig. 1, disrupts the path's functionality.

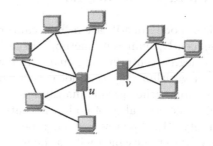

Fig. 1. Sample network represented as a graph

Numerous studies [7, 36, 49] have proposed various methods for the identification of critical nodes in networks. However, some measures may prove ineffective when dealing with specific network vulnerabilities. This becomes particularly evident when prioritizing the prevention of network fragmentation, as demonstrated in Fig. 1, where the compromise of node u may fragment the network into two disconnected networks. Furthermore, some methods [8,44] rely on solving computationally expensive and intractable NP-hard optimization problems, which may not be practical or feasible in all scenarios.

Graph-based features like centrality measures [16] have proven to be effective in critical node identification. In this research we utilize one of these centrality measures – the closeness centrality measure – to identify and analyze critical nodes. Nevertheless, our methodology can be extended to encompass other measures like degree and betweenness centrality.

Our approach distinguishes itself through the innovative use of information acquired during the computation of node centrality measures (or scores), enabling us to compute the probability of a critical node becoming the target of an attack. At its core, our strategy is centered on the idea that enhancing the security of critical nodes significantly strengthens network robustness and reduces vulnerability to cyberattacks. As a result, we propose a non-linear optimization problem, aiming to minimize the likelihood of a network attack subject to a budget constraint. The solution to this optimization problem yields an optimal allocation of resources, measured in terms of cost, to each critical node, thereby effectively mitigating the risk of a cyberattack.

The remainder of this paper is organized as follows. Section 2 presents an overview of related work in this domain and explores its relevant applications. In Sect. 3, we introduce our network model and detail the method we employ for identifying critical nodes. In Sect. 4, we delve into various insights and properties uncovered within our network, including its susceptibility to cyberattacks and robustness. Additionally, we present our optimization model and its solution. Finally, Sect. 5 summarizes our work and outlines potential directions for future research.

2   Related Work and Applications

The problem of critical node identification has captured the attention of researchers [58]. Consequently, numerous methods have been devised, including but not limited to degree centrality [2], mapping entropy centrality [46], the collective influence algorithm [45], semi-local centrality [13], coreness [9], and H-index [42]. These approaches predominantly rely on node-local neighbor information to determine the criticality of nodes. Eccentricity [24], information indices [51], Katz centrality [31], closeness centrality, betweenness centrality [47], and subgraph centrality [18] are designed with a focus on the number of paths for communication. In addition to the methods mentioned above, there are also numerous iterative refinement algorithms that take into account not only the number of neighbors but also the significance of these neighbors. Representative algorithms include PageRank [12], HITs [35], and eigenvector centrality [11].

The problem of critical node identification has also gained substantial attention across diverse fields. While our primary focus in this paper centers on communication networks, network vulnerability, and security issues, it is crucial to acknowledge the broad relevance and importance of this problem in other domains. The critical node identification problem has found applications in fields such as transportation [23], social networks [43], biology [38], public health [3], fraud detection [40], intrusion detection [15], image processing [59], and astronomical data analysis [1,37].

In the realm of network vulnerability assessment, much of the existing research [17,22,34] has focused on centrality measurements, including degree, betweenness, closeness centralities, and average shortest path length, as these metrics have been instrumental in understanding network structures. For instance, in [32] researchers showcased the practicality of centrality measures in identifying nodes involved in malware distribution, offering insights into distinguishing critical (e.g., malicious) nodes from non-critical (e.g., benign) nodes.

Research efforts [55,60] have employed graph measurements to identify critical nodes by primarily relying on metrics such as average similarity and global clustering coefficients. They found that critical nodes tend to exhibit higher similarity and tighter clustering, providing valuable insights into network vulnerability. Notably, these studies did not comprehensively analyze the wide range of structural differences that may exist between critical and non-critical nodes.

In telecommunication networks, the identification of critical nodes holds paramount importance, serving both defensive and offensive purposes. These nodes play a dual role, either as essential components to maintain the functionality of a communication network or as prime targets for disruption in adversarial contexts. Consider the case of terrorist and insurgent networks, where the primary objective is to serve the communication channels by strategically removing critical nodes to disable terrorist networks [6]. Similarly, in the context of wireless communication networks, the Wireless Network Jamming Problem is formulated as a critical node identification challenge [14]. In this scenario, the goal is to pinpoint critical nodes that, when jammed, effectively neutralize an adversary's wireless communication network.

In sensor networks [28,50], beyond its role in anomaly detection, critical node identification plays a crucial role in optimizing energy utilization and extending the operational lifespan of these nodes [25]. Critical nodes, often traversed by many shortest paths, can experience faster energy depletion, potentially leading to network fragmentation.

In decentralized systems such as peer-to-peer and adhoc networks [26,30], a major weakness is network disconnectivity. These systems typically have a weak topology that can be easily fragmented by targeting critical nodes, which maintain the network's entire connectivity. Therefore, the identification of these critical nodes is essential for designing such networks to be robust and secure.

Research in [29,56] delved into the study of malware distribution networks using centrality metrics and empirical approaches [33]. Others [19,20,37,39] have framed the critical node identification problem as an optimization challenge. This entails the identification of a set of nodes whose removal would significantly degrade network connectivity based on predefined metrics. While these problems can be computationally demanding, researchers have explored various approaches, including dynamic programming and integer linear programming, to seek exact solutions. Moreover, there have been efforts to develop approximate solutions with performance guarantees, employing heuristic algorithms and polynomial-time approximation algorithms [5,10,53].

The study in [52] applied complex network theory to assess the importance of nodes within communication networks. The proposed approach identified critical nodes by combining various characteristic of the communication network. More precisely, it established node contribution metrics by considering both a node receiving information and a node providing information to its neighboring nodes. Node importance was subsequently computed by integrating these metrics with the node's own attributes.

3   Model Description

We model a communication network as an undirected and unweighted graph denoted by $G = (V, E)$, as illustrated in Fig. 2. In this representation, V corresponds to the collection of nodes, encompassing diverse components such as routers, switches, and similar elements, with $|V| = n$. Furthermore, E represents the set of edges or links $(E \subset V \times V)$, with $|E| = m$.

In such networks, it's essential to acknowledge that nodes exhibit varying degrees of importance. For instance, certain nodes, such as routers, assume pivotal roles, as their removal could lead to network disconnection and operational disruptions. On the other hand, strategically or centrally located nodes often store significant data, such as servers. In contrast, peripheral or end nodes generally have less overall significance in network operations.

In the context of communication networks, the significance of a node can be assessed by several factors. A node is considered important if it plays a pivotal role in the network's flow dynamics. This can manifest in various ways: a node may be important if it generates, receives, or facilitates a big volume of data flows. Additionally, a node can be important if it resides along critical paths through which network traffic is routed. In essence, the importance of a node is often measured by its role in efficiently managing data flows within the network.

The concept of centrality serves as a foundational tool in graph analysis, helping in the identification of important and key nodes within a network. Various centrality measures exist, each illuminating a distinct facet of node importance [16]. In our study, we focus on nodes that excel in efficiently disseminating information throughout the network. These nodes are known to have high closeness scores, and consequently, we employ this centrality measure to identify critical nodes and conduct our analysis. For completeness, in the following we will provide a brief description of the closeness centrality method.

3.1   Closeness Centrality

Node closeness centrality [16,21,57] is a metric that assesses how closely connected a specific node is to all other nodes within a network. It is formally defined as the reciprocal of the sum of the shortest path distances between the given node and all other nodes in the network. The closeness centrality of a node i, denoted by C_i, can be calculated using the formula: $C_i = (n-1)/\sum_i^j d_{ij}$. Here, d_{ij} represents the shortest path distance between nodes i and j.

This measure proves invaluable in understanding the network's overall structure. Nodes with high closeness centrality often serve as vital hubs for information flow. They can facilitate the dissemination of information, including potential threats like malware, thereby influencing other nodes and the network as a whole. Conversely, nodes with high closeness centrality scores may indicate potential problems within the network. These issues could range from communication link failures to the presence of isolating malware, resulting in the node being disconnected from the broader network.

Closeness centrality offers a notable advantage as it is computed based on a global view of the network, making it responsive to network changes such as nodes joining or leaving. However, this advantage comes at the expense of computational complexity, particularly for large and intricate networks, with a computational complexity of $O(mn)$ [16]. Additionally, the requirement of node reachability in closeness centrality estimation introduces limitations, making it unsuitable for networks with disconnected components or networks characterized by small diameters [48].

3.2 Critical Node Identification

We consider a network comprising 17 nodes and 38 edges, graphically represented in Fig. 2, for our node identification and analysis. Figure 3 presents the calculated importance scores – closeness centrality scores C_i for all nodes i within the network, showcased in descending order.

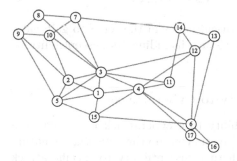

Fig. 2. Network represented as a graph.

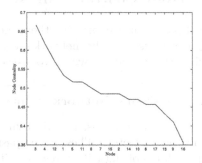

Fig. 3. Nodes ranked by centrality.

For our analysis and to mitigate computational complexity, we will narrow our focus to the subset of critical nodes with importance scores exceeding a specified threshold $\beta \geq 0$ ($C_i \geq \beta$). Table 1 presents the nodes that meet our criteria with $\beta = 0.5$. Let S denote the collection of these critical nodes, with $|S| = k$. These nodes are highlighted in blue color in Fig. 4.

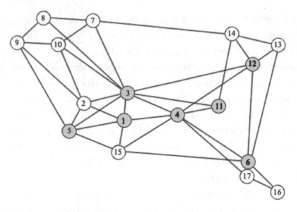

Table 1. Most critical nodes $(\beta = 0.5)$.

Node i	Centrality measure C_i
3	0.666
4	0.615
12	0.571
1	0.533
5	0.516
11	0.516
6	0.500

Fig. 4. Most critical nodes (in blue). (Color figure online)

Our numerical results in Table 1 show that node 3 emerges as the most critical node in the network, closely followed by node 4. While we have presented only a subset of the network's critical nodes, our computations reveal that the least critical node is node 16, with an importance score of 0.355. The network's topology, illustrated in Fig. 4, aligns with our findings. Node 16 is located on the periphery of the network, distant from the majority of other nodes, hence its low importance score. Conversely, node 3 holds a central position within the network, enabling it to be in close proximity to a greater number of nodes, thus underscoring its high importance score.

4   Node Centrality in Network Assessment

In this section, we examine the k most critical nodes in the set S to gain valuable insights into the network's characteristics, including its susceptibility to cyberattacks and its overall robustness.

4.1   Assessing Network Vulnerability to Cyberattacks

In practical scenarios, a node's vulnerability to a cyberattack is shaped by a multitude of factors. These encompass its position within the network, the nature and volume of data it stores, and its inherent susceptibility to specific attack vectors. The interconnected nature of network nodes further complicates the assessment of vulnerability, as the likelihood of an attack hinges on the collective characteristics of all nodes, particularly those in immediate proximity.

The node's location within the network not only determines its exposure to potential threats but also shapes its role in the overall security posture of the network. Nodes situated at critical locations may attract attackers due to their capacity to disrupt data flow or compromise the network's integrity. Furthermore, the type and volume of data a node handles are pivotal factors. Nodes that store sensitive or valuable information become high-priority targets, increasing their vulnerability. Additionally, a node's susceptibility extends to specific attack methods. Vulnerabilities, such as unpatched software or weak security protocols, can increase the risk of exploitation by cybercriminals. Finally, nodes in close proximity can trigger a cascading effect in terms of security. The compromise of one node may open pathways to adjacent nodes, potentially amplifying the impact of an attack.

In our analysis, we operate under the assumption that these factors and node attributes are reflected in the closeness centrality measure, C_i of each node i. By leveraging the centrality measures of critical nodes, we enhance our capability to approximate and evaluate the network's susceptibility to attacks. This, in turn, empowers us to reinforce our defensive strategies and efficiently enhance network security. To achieve this, we've standardized the centrality scores, denoted as p_i for nodes $i \in S$, to ensure that the sum of centralities across all critical nodes equals one. This normalization process enables us to introduce the following interpretation of the p_i values.

Interpretation: *The normalized centrality measure p_i for node i represents its vulnerability to cyberattacks, with higher values indicating a greater likelihood of being targeted. Mathematically, p_i is defined as: $p_i = C_i / \sum_{j \in S} C_j$, and can be interpreted as the probability of attack of node i.*

The aforementioned interpretation introduces an innovative concept that establishes a correlation between a node's centrality measure and its susceptibility to attacks. Table 2 provides the probability p_i of the k critical nodes (*i.e.* the probability of an attack on node 3 is $p_3 = 0.170$).

Table 2. Centrality scores (C_i) for critical nodes and their corresponding attack probabilities (p_i)

Node i	Centrality measure C_i	Probability of attack p_i
3	0.666	0.170
4	0.615	0.157
12	0.571	0.146
1	0.533	0.136
5	0.516	0.132
11	0.516	0.132
6	0.500	0.128

4.2   Evaluating Network Robustness

In the realm of network security, the robustness of a network is intricately dependant on the security and stability of its individual nodes. To quantify it, we denote the network robustness as R and formally define it as follows:

Definition 1. *The network robustness, R, quantifies the probability that the identified k critical nodes in the network remain secure. It is defined as a function of the node-level probability of compromise, p_i, associated with each node $i \in S$:*

$$R = \prod_{i \in S, |S| = k} (1 - p_i).$$

In essence, R is a metric that provides a quantitative measure of a network's ability to withstand a potential disruptions, including cyberattacks by accounting for the combined resilience of its critical nodes. It serves as a valuable tool for network administrators, cybersecurity experts, and decision-makers, offering insights into the network's vulnerability and resilience.

Based on our numerical results, as shown in Table 2, the network's robustness is computed to be 33.95%. This value indicates a positive yet relatively low level of robustness. It's worth noting that the network's overall robustness, considering all 17 nodes, is computed at 35.56%, which is 4.75% higher than when considering only the k most critical nodes. This difference, while noteworthy, may be negligible when factoring in the computational and investment costs associated with securing low-importance nodes.

4.3   Enhancing Network Security Through Critical Node Hardening

Given the low network robustness discussed in the previous subsection, our primary objective here is to strengthen network security and proactively defend against emerging cyber threats. Achieving this entails the comprehensive implementation of protective measures at critical nodes, potentially encompassing technologies such as firewalls, Intrusion Detection Systems, Virtual Private Networks, access control systems, antivirus software, Data Loss Prevention systems, and the establishment of well-defined security policies.

This process involves dedicated investments in enhancing the security of individual nodes, denoted as $s_i, i \in S$. Striking the right balance between security costs and the probability of potential attacks is pivotal in devising effective strategies to safeguard these critical nodes.

It's worth noting that there exists an inverse correlation between the probability of a critical node being compromised and the level of security investment it receives. We leverage this relationship to frame the security of critical nodes as a non-linear optimization problem. Our objective is to identify the optimal values of s_i for nodes $i \in S$ that minimize the probability of critical node compromise (p_i) within a specified budget. Conversely, the objective can be restated as maximizing network robustness R.

For the sake of clarity, we express p_i as a function of s_i and denote it as $p_i(s_i)$. We assume that $p_i(s_i) = p_i(0)e^{-\alpha s_i}$, where α is a predefined positive constant, and $p_i(0)$ represents the probability of node i being compromised when $s_i = 0$. *i.e.* the probability of a node attack before any security investment. α can be interpreted as the sensitivity of the probability of attack to the security investment level. It's worth noting that alternative functions, such as polynomial functions, can be considered.

The non-linear optimization model, as outlined below (Eqs. 1 through 4), is designed to allocate a budget of B dollars (as described in Eq. 3) towards strengthening the security of the k critical nodes.

$$\max R = \prod_{i \in S}(1 - p_i(s_i)). \tag{1}$$

$$\text{s.t.} \quad 0 \le p_i(s_i) = p_i(0)e^{-\alpha s_i} \le 1 \tag{2}$$

$$\sum_{i \in S} s_i \le B \tag{3}$$

$$s_i \ge 0, \forall i \in \{1, \ldots, k\}. \tag{4}$$

Proposition 1. *The function R is convex in* $\mathbf{s} = (s_1, \ldots s_k)$.

Proof. In this proof we begin by applying the logarithm (*log*) of the product terms of R, resulting in $log(\prod_{i \in S}(1 - p_i(s_i)))$. This simplifies to: $\sum_{i \in S} log(1 - p_i(s_i))$. Recognizing that $(1 - p_i(s_i))$ is convex in s_i, the *log* preserves convexity, and since the sum of convex functions is convex, we conclude that the function R is convex.

We solved the optimization problem using *fmincon* function in MATLAB for the example network illustrated in Fig. 2 and the critical nodes and centrality values given in Table 2. The results presented in Table 3 show that as the budget increases (from $B = 0$ to $B = 500$), the probabilities of an attack decreases. For instance, with a budget of zero, the probability of an attack on node 3 is 0.170, however, as the budget increases to 300, the probability drops to zero. This decrease in attack probability leads to an increase in network robustness, as illustrated in Table 3 and Fig. 5. When the budget is zero, the network's robustness is $R = 33.9\%$. However, with a budget of 300, the network's robustness increases to $R = 48.6\%$, representing a notable increase of 14.7%. Figure 5 shows that there is a positive correlation between the network robustness and the budget.

Table 3. Optimal values of the attack probability (p_i^*) for critical node attacks and the network's optimal robustness (R^*) across different budget levels.

Node	$B = 0$	$B = 100$	$B = 200$	$B = 300$	$B = 400$	$B = 500$
i	p_i^*	p_i^*	p_i^*	p_i^*	p_i^*	p_i^*
3	0.170	0.043	0	0	0	0
4	0.157	0.117	0.040	0	0	0
12	0.146	0.146	0.109	0.037	0.002	0
1	0.136	0.136	0.136	0.102	0.068	0.034
5	0.132	0.132	0.132	0.132	0.099	0.066
11	0.132	0.132	0.132	0.132	0.099	0.066
6	0.128	0.128	0.128	0.128	0.126	0.095
$R^*(\%)$	33.9	41.0	48.6	56.9	66.0	76.2

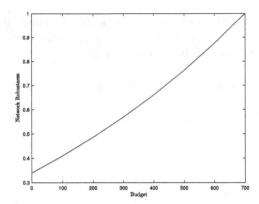

Fig. 5. Network robustness as a function of the budget.

Note that with an investment of $s_3^* = \$74.55$ out of the available \$100 budget, we managed to significantly reduce the attack probability on node 3 by 74.55% (which we denote as $p_{\downarrow}(\%)$ in Table 4). These values are highlighted in bold in Table 4.

It's worth highlighting that a substantial portion of the budget is strategically assigned to critical nodes based on their centrality scores. In scenarios with limited budgets (e.g., $B = 100$), a majority of the resources are directed towards nodes 3 and 4. Conversely, in scenarios with a more generous budget (e.g., $B = 500$), substantial resources are allocated to nodes 3 and 4, effectively reducing the attack probability to zero, while the remaining funds are distributed among less critical nodes.

Table 4. Optimal budget allocation among the critical nodes which led to a reduction of the probability of an attack ($p_{\downarrow}(\%)$).

Node	$B = 0$	$B = 100$		$B = 200$		$B = 300$		$B = 400$		$B = 500$	
i	p_i	s_i	$p_{\downarrow}(\%)$	s_i	$p_{\downarrow}(\%)$	s_i	$p_{\downarrow}(\%)$	s_i	$p_{\downarrow}(\%)$	s_i	$p_{\downarrow}(\%)$
3	**0.170**	**74.55**	**74.54**	100	100	100	100	100	100	100	100
4	0.157	25.44	25.41	74.42	74.39	100	100	100	100	100	100
12	0.146	0	0	25.58	25.58	74.85	74.83	98.54	98.56	100	100
1	0.136	0	0	0	0	24.89	24.89	50.19	50.22	75.08	75.11
5	0.132	0	0	0	0	0.13	0.08	24.96	24.92	49.69	49.70
11	0.132	0	0	0	0	0.13	0.08	24.96	24.92	49.69	49.70
6	0.128	0	0	0	0	0	0	1.36	1.41	25.54	25.55

5   Conclusion

In this research, we present an innovative approach that harnesses insights derived from the computation of node centrality scores using the closeness centrality measure. The novelty of our work is rooted in the transformation of centrality measures into attack probability, providing a distinctive viewpoint on network security for resource allocation and the enhancement of overall network resilience. Additionally, we present a closed-form expression for network robustness, establishing its correlation with node features such as importance scores and attack probabilities. We further formulate the network as a non-linear optimization problem, accounting for budget constraints. Through this optimization framework, we determine the optimal allocation of resources, to reduce the probability of a cyberattack on critical nodes. Our findings underscore the positive impact of resource allocation to critical nodes, resulting in enhanced network robustness and a reduced likelihood of an attack.

Our future work will focus on validating our findings in larger networks and enhancing our critical node identification techniques. We believe that these ongoing efforts will continue to bolster the practical applicability and effectiveness of our approach, thereby contributing to the safeguarding of critical infrastructure and the reinforcement of cybersecurity measures in communication network.

References

1. Ahmed, M., Naser Mahmood, A., Hu, J.: A survey of network anomaly detection techniques. J. Netw. Comput. Appl. **60**, 19–31 (2016)
2. Albert, R., Jeong, H., Barabási, A.L.: Diameter of the world-wide web. Nature **401**(6749), 130–131 (1999)
3. Alozie, G.U., Arulselvan, A., Akartunalı, K., Pasiliao, E.L., Jr.: A heuristic approach for the distance-based critical node detection problem in complex networks. J. Oper. Res. Soc. **73**(6), 1347–1361 (2022)
4. Amini, M.H., Arasteh, H., Siano, P.: Sustainable smart cities through the lens of complex interdependent infrastructures: panorama and state-of-the-art. In: Amini, M.H., Boroojeni, K.G., Iyengar, S.S., Pardalos, P.M., Blaabjerg, F., Madni, A.M. (eds.) Sustainable Interdependent Networks II. SSDC, vol. 186, pp. 45–68. Springer, Cham (2019). https://doi.org/10.1007/978-3-319-98923-5_3
5. Aringhieri, R., Grosso, A., Hosteins, P., Scatamacchia, R.: A general evolutionary framework for different classes of critical node problems. Eng. Appl. Artif. Intell. **55**, 128–145 (2016)
6. Arulselvan, A.: Network model for disaster management. Ph.D. thesis, University of Florida Gainesville (2009)
7. Arulselvan, A., Commander, C., Elefteriadou, L., Pardalos, P.: Detecting critical nodes in sparse graphs. Comput. Oper. Res. **36**, 2193–2200 (2009). https://doi.org/10.1016/j.cor.2008.08.016
8. Arulselvan, A., Commander, C.W., Elefteriadou, L., Pardalos, P.M.: Detecting critical nodes in sparse graphs. Comput. Oper. Res. **36**(7), 2193–2200 (2009)
9. Bae, J., Kim, S.: Identifying and ranking influential spreaders in complex networks by neighborhood coreness. Phys. A **395**, 549–559 (2014)

10. Berger, A., Grigoriev, A., van der Zwaan, R.: Complexity and approximability of the k-way vertex cut. Networks **63**(2), 170–178 (2014)
11. Bonacich, P.: Factoring and weighting approaches to status scores and clique identification. J. Math. Sociol. **2**(1), 113–120 (1972)
12. Brin, S., Page, L.: The anatomy of a large-scale hypertextual web search engine. Comput. Netw. ISDN Syst. **30**(1–7), 107–117 (1998)
13. Chen, D., Lü, L., Shang, M.S., Zhang, Y.C., Zhou, T.: Identifying influential nodes in complex networks. Phys. A **391**(4), 1777–1787 (2012)
14. Commander, C.W., Pardalos, P.M., Ryabchenko, V., Uryasev, S., Zrazhevsky, G.: The wireless network jamming problem. J. Comb. Optim. **14**, 481–498 (2007)
15. Dang, F., Zhao, X., Yan, L., Wu, K., Li, S.: Research on network intrusion response method based on Bayesian attack graph, pp. 639–645 (2023)
16. Das, K., Samanta, S., Pal, M.: Study on centrality measures in social networks: a survey. Soc. Netw. Anal. Min. (13) (2018)
17. Devkota, P., Danzi, M.C., Wuchty, S.: Beyond degree and betweenness centrality: alternative topological measures to predict viral targets. PLoS ONE **13**(5), e0197595 (2018)
18. Estrada, E., Rodriguez-Velazquez, J.A.: Subgraph centrality in complex networks. Phys. Rev. E **71**(5), 056103 (2005)
19. Faramondi, L., Oliva, G., Pascucci, F., Panzieri, S., Setola, R.: Critical node detection based on attacker preferences, pp. 773–778 (2016)
20. Faramondi, L., Oliva, G., Setola, R., Pascucci, F., Esposito Amideo, A., Scaparra, M.P.: Performance analysis of single and multi-objective approaches for the critical node detection problem, pp. 315–324 (2017)
21. Fernandes, J.M., Suzuki, G.M., Zhao, L., Carneiro, M.G.: Data classification via centrality measures of complex networks, pp. 1–8 (2023). https://doi.org/10.1109/IJCNN54540.2023.10192048
22. Freeman, L.: Centrality in social networks conceptual clarification. Soc. Netw. **1**, 215 (1979)
23. Gupta, B.B., Gaurav, A., Marín, E.C., Alhalabi, W.: Novel graph-based machine learning technique to secure smart vehicles in intelligent transportation systems. IEEE Trans. Intell. Transp. Syst. **24**(8), 8483–8491 (2023). https://doi.org/10.1109/TITS.2022.3174333
24. Hage, P., Harary, F.: Eccentricity and centrality in networks. Soc. Netw. **17**(1), 57–63 (1995)
25. Hamouda, E., Mitton, N., Pavkovic, B., Simplot-Ryl, D.: Energy-aware georouting with guaranteed delivery in wireless sensor networks with obstacles. Int. J. Wirel. Inf. Netw. **16**, 142–153 (2009)
26. Hamouda, E., Mitton, N., Simplot-Ryl, D.: Energy efficient mobile routing in actuator and sensor networks with connectivity preservation. In: Frey, H., Li, X., Ruehrup, S. (eds.) ADHOC-NOW 2011. LNCS, vol. 6811, pp. 15–28. Springer, Heidelberg (2011). https://doi.org/10.1007/978-3-642-22450-8_2
27. Hao, Y.H., Han, J.H., Lin, Y., Liu, L.: Vulnerability of complex networks under three-level-tree attacks. Phys. A **462**, 674–683 (2016)
28. Imran, M., Alnuem, M.A., Fayed, M.S., Alamri, A.: Localized algorithm for segregation of critical non-critical nodes in mobile ad hoc and sensor networks. Procedia Comput. Sci. **19**, 1167–1172 (2013). https://doi.org/10.1016/j.procs.2013.06.166. https://www.sciencedirect.com/science/article/pii/S1877050913007746. The 4th International Conference on Ambient Systems, Networks and Technologies (ANT 2013), the 3rd International Conference on Sustainable Energy Information Technology (SEIT-2013)

29. Invernizzi, L., et al.: Nazca: detecting malware distribution in large-scale networks (2014)
30. Jain, A., Reddy, B.: Node centrality in wireless sensor networks: importance, applications and advances. In: Proceedings of the 2013 3rd IEEE International Advance Computing Conference, IACC 2013, pp. 127–131 (2013). https://doi.org/10.1109/IAdCC.2013.6514207
31. Katz, L.: A new status index derived from sociometric analysis. Psychometrika 18(1), 39–43 (1953)
32. Kim, S.: Anatomy on malware distribution networks. IEEE Access 8, 73919–73930 (2020). https://doi.org/10.1109/ACCESS.2020.2985990
33. Kim, S., Kim, J., Kang, B.B.: Malicious URL protection based on attackers' habitual behavioral analysis. Comput. Secur. 77, 790–806 (2018)
34. Kivimäki, I., Lebichot, B., Saramäki, J., Saerens, M.: Two betweenness centrality measures based on randomized shortest paths. Sci. Rep. 6(1), 1–15 (2016)
35. Kleinberg, J.M.: Authoritative sources in a hyperlinked environment. J. ACM (JACM) 46(5), 604–632 (1999)
36. Lalou, M., Tahraoui, M.A., Kheddouci, H.: The critical node detection problem in networks: a survey. Comput. Sci. Rev. 28, 92–117 (2018)
37. Lalou, M., Tahraoui, M.A., Kheddouci, H.: The critical node detection problem in networks: a survey. Comput. Sci. Rev. 28, 92–117 (2018). https://doi.org/10.1016/j.cosrev.2018.02.002. https://www.sciencedirect.com/science/article/pii/S1574013716302416
38. Liu, X., Hong, Z., Liu, J., Lin, Y., et al.: Computational methods for identifying the critical nodes in biological networks. Brief. Bioinform. 21, 486–497 (2020)
39. Lozano, M., Garcia-Martinez, C., Rodriguez, F.J., Trujillo, H.M.: Optimizing network attacks by artificial bee colony. Inf. Sci. 377, 30–50 (2017)
40. Lu, K., Fang, X., Fang, N.: PN-BBN: a petri net-based Bayesian network for anomalous behavior detection. Mathematics 10(20), 3790 (2022)
41. Lü, L., Chen, D., Ren, X.L., Zhang, Q.M., Zhang, Y.C., Zhou, T.: Vital nodes identification in complex networks. Phys. Rep. 650, 1–63 (2016)
42. Lü, L., Zhou, T., Zhang, Q.M., Stanley, H.E.: The h-index of a network node and its relation to degree and coreness. Nat. Commun. 7(1), 10168 (2016)
43. Mazlumi, S.H.H., Kermani, M.A.M.: Investigating the structure of the internet of things patent network using social network analysis. IEEE Internet Things J. 9(15), 13458–13469 (2022). https://doi.org/10.1109/JIOT.2022.3142191
44. Lalou, M., Tahraoui, M.A., Kheddouci, H.: The critical node detection problem in networks: a survey. Comput. Sci. Rev. 28, 92–117 (2018). https://doi.org/10.1016/j.cosrev.2018.02.002
45. Morone, F., Makse, H.A.: Influence maximization in complex networks through optimal percolation. Nature 524(7563), 65–68 (2015)
46. Nie, T., Guo, Z., Zhao, K., Lu, Z.M.: Using mapping entropy to identify node centrality in complex networks. Phys. A 453, 290–297 (2016)
47. Sabidussi, G.: The centrality index of a graph. Psychometrika 31(4), 581–603 (1966)
48. Sariyüce, A.E., Kaya, K., Saule, E., Çatalyiirek, Ü.V.: Incremental algorithms for closeness centrality, pp. 487–492 (2013)
49. Shen, Y., Nguyen, N., Xuan, Y., Thai, M.: On the discovery of critical links and nodes for assessing network vulnerability. IEEE/ACM Trans. Netw. 21, 963–973 (2013). https://doi.org/10.1109/TNET.2012.2215882
50. Shukla, S.: Angle based critical nodes detection (ABCND) for reliable industrial wireless sensor networks. Wireless Pers. Commun. 130(2), 757–775 (2023)

51. Stephenson, K., Zelen, M.: Rethinking centrality: methods and examples. Soc. Netw. **11**(1), 1–37 (1989)

52. Tian, G., Yang, X., Li, Y., Yang, Z., Chen, G.: Hybrid weighted communication network node importance evaluation method. Front. Phys. **11** (2023). https://doi.org/10.3389/fphy.2023.1133250. https://www.frontiersin.org/articles/10.3389/fphy.2023.1133250

53. Ventresca, M., Aleman, D.: A derandomized approximation algorithm for the critical node detection problem. Comput. Oper. Res. **43**, 261–270 (2014)

54. Wandelt, S., Lin, W., Sun, X., Zanin, M.: From random failures to targeted attacks in network dismantling. Reliab. Eng. Syst. Saf. **218**, 108146 (2021). https://doi.org/10.1016/j.ress.2021.108146

55. Wang, B., Jia, J., Zhang, L., Gong, N.Z.: Structure-based Sybil detection in social networks via local rule-based propagation. IEEE Trans. Netw. Sci. Eng. **6**, 523–537 (2018)

56. Yan, G., Chen, G., Eidenbenz, S.J., Li, N.: Malware propagation in online social networks: nature, dynamics, and defense implications (2011)

57. Yen, C.C., Yeh, M.Y., Chen, M.S.: An efficient approach to updating closeness centrality and average path length in dynamic networks, pp. 867–876 (2013). https://doi.org/10.1109/ICDM.2013.135

58. Yi-Run, R., Song-Yang, L., Yan-Dong, X., Jun-De, W., Liang, B.: Identifying influence of nodes in complex networks with coreness centrality: decreasing the impact of densely local connection. Chin. Phys. Lett. **33**(2), 028901 (2016)

59. Zhang, S., Yu, H., et al.: Modeling and simulation of tennis serve image path correction optimization based on deep learning. Wirel. Commun. Mob. Comput. **2022** (2022)

60. Zheng, H., et al.: Smoke screener or straight shooter: detecting elite Sybil attacks in user-review social networks. arXiv:abs/1709.06916 (2017)

Strategic Decision-Making in Trauma Systems

Eva K. Lee[1,2,3(✉)] ⓘ, A Oguzhan Ozlu[2,4], Taylor J. Leonard[2,5] ⓘ, Michael Wright[6],
and Daniel Wood[7]

[1] The Data and Analytics Innovation Institute, Atlanta, GA 30309, USA
`evalee-gatech@pm.me`
[2] Georgia Institute of Technology, Atlanta, GA 30322, USA
[3] Accuhealth Technologies, Atlanta, GA 30310, USA
[4] T-Mobile, Atlanta, GA, USA
[5] The United States Department of Air Force, Pentagon, Washinton D.C 20330, USA
[6] Grady Health Systems, Atlanta, GA 30303, USA
[7] Emory University Hospital Midtown, Atlanta, GA 30308, USA

Abstract. Trauma care and trauma systems are vital community assets. A trauma
system manages the treatment of severely injured people and spans the full spectrum of prevention and emergency care to recovery and rehabilitation. While most
trauma-related studies concern the operational and tactical levels of trauma care
and trauma systems, little attention has been given to a strategic level (top-down)
approach to a trauma system from a financial and investment perspective. In this
paper, we analyze a statewide trauma system and model it as a network of trauma
centers, hospitals and emergency medical services (EMS). We develop a theoretical model and a general-purpose computational framework that facilitates the
allocation and utilization of resources by the statewide trauma system. Given a
local trauma network profile, injury distribution and resource requests, the modeling and computational framework enable the creation of the set of all feasible
and Pareto-efficient portfolios where limited funding is allocated across numerous
resource requests from trauma centers, hospitals and EMS providers. Using the
computational framework, decision-makers can quantitatively analyze the impact
of each feasible portfolio on the trauma system's performance measures via the
Trauma System Simulator. Sensitivity analysis can be conducted to determine the
best decision or policy for transporting and transferring patients and to observe
how changes in system inputs affect the return on investment (ROI) and resource
utilization. Using the trauma system data in Georgia, our findings confirm that such
dynamic and strategic resource-allocation analyses empower decision-makers to
make informed decisions that benefit the entire trauma network. The design is
a top-down approach at a strategic level that simultaneously uses tactical-level
decisions to evaluate several strategies to improve the trauma system.

Keywords: Trauma System · System and Network Modeling · System
Simulation and Optimization · Pareto-Efficient Frontier · Trauma Centers ·
Emergency Medical Services · Strategic Decisions · Resource Allocation ·
Trauma Patient Transport and Transfer · Trauma Outcome Metrics · System
Performance Measures · Return on Investment

H. Moosaei et al. (Eds.): DIS 2023, LNCS 14321, pp. 131–158, 2024.
https://doi.org/10.1007/978-3-031-50320-7_10

1 Introduction

Background. Trauma systems are critical for ensuring that individuals who experience traumatic injuries receive the best possible care. They facilitate a rapid and coordinated response, provide specialized care, and contribute to the overall health and safety of communities. These systems are essential for reducing mortality, preventing disability and improving the quality of care for trauma patients.

The study of trauma care systems and trauma policy development began after the Vietnam War [1] with the creation of trauma centers and regional trauma systems. Numerous studies have demonstrated that the implementation of a statewide trauma system reduces the frequency of hospitalizations and death [2–8]. The literature covers a broad set of issues to improve the quality of trauma care. One issue is the impact of transport time on patient survival rates, with findings suggesting that transport time from the scene of the incident to the hospital does not impact survival rates. However, the time it takes for emergency responders to arrive on the scene once they have received the call does have an impact [7, 9–12]. Other patient-centered studies focus on treatment and intervention [13–16], as well as injury evaluation methods [17–21]. There are also retrospective reviews of trauma patient data and related statistics on factors that affect the mortality and morbidity of trauma patients [22, 23]. As most trauma-related studies concern the operational and tactical levels of trauma care and trauma systems, little attention has been given to a strategic level (top-down) approach to a trauma system from a financial and investment perspective.

In this paper, we analyze the Georgia state trauma system and model it as a network of trauma facilities, hospitals, and emergency medical services (EMS). Our goal is to design a long-term development model of the network by considering the demands of each component in the network and the strategies to maximize the quality of patient care and patient outcomes, without creating a financial burden on the system's trauma centers, EMS and hospitals. This is a critical but thus far untouched task.

There are four major perspectives to address when seeking to improve patient outcomes: (a) the patient's perspective, (b) the EMS perspective, (c) the hospital/trauma facility perspective and (d) the policy/systems perspective. Research on patients often addresses how the survival, mortality and morbidity rates are affected by the characteristics of patients [24–29] and clinical decision-making [30–32]. Studies that concern emergency medicine assess the impact of emergency response time on patient outcomes [30, 33, 34] and the effect of pre-hospital trauma care on trauma patient survival rates [19]. Other works focus on developing new rules to predict emergency interventions that will improve triage effectiveness and efficiency [6, 16, 18]. Patient transportation strategies to trauma facilities or emergency departments [23, 35–37] as well as trauma system effectiveness are also studied extensively.

Trauma Centers and Research Gaps. This project aims to improve the Georgia trauma care system and support the Georgia Trauma Care Network Commission's (GTCNC) five-year strategic plan to address existing deficiencies and recommend future developments [38]. The methodologies, results and implications of the work are relevant not only to Georgia but to other states.

Trauma centers are specialized healthcare facilities equipped to provide comprehensive emergency medical services for patients with severe injuries. Trauma centers are typically identified with a designation of Level I – Level IV, depending on their resources, capabilities, and ability to provide specialized care [39]. The levels may vary slightly from one country or region to another, but the four levels of trauma centers are as follows:

Level I trauma centers are the highest level of trauma care facilities. They are typically major teaching hospitals or medical centers with 24/7 availability of highly specialized surgeons, physicians and support staff. Level I trauma centers have a full range of medical specialties, including emergency medicine, neurosurgery, orthopedic surgery and anesthesiology, among others. These centers have the capacity to provide immediate resuscitation, surgical intervention, intensive care and rehabilitation services for trauma patients. Level I trauma centers also play a vital role in research and education related to trauma care.

Level II trauma centers are also comprehensive facilities, but they may have slightly fewer resources compared to Level I centers. They have the capability to provide initial evaluation, stabilization and surgical intervention for trauma patients. Level II centers often have access to specialized surgeons and physicians but may not offer the full spectrum of medical specialties found in Level I centers. However, they can provide immediate life-saving interventions and transfer patients to Level I facilities if needed.

Level III trauma centers are typically community or regional hospitals that have the resources to provide prompt assessment, resuscitation and stabilization of trauma patients. These centers have emergency departments capable of providing initial trauma care, but their resources may be limited compared to Level I and II centers. Level III centers usually have partnerships or transfer agreements with higher-level trauma centers to transfer patients requiring specialized care.

Level IV trauma centers are generally small, rural hospitals that provide initial evaluation, stabilization and transfer of trauma patients to higher-level trauma centers. They may have limited resources and capabilities compared to higher-level centers. Level IV centers focus on stabilizing patients before they can be transferred to a Level I, II, or III trauma center, which can provide the necessary advanced care.

Discrete time simulation has been used extensively in modeling emergency medical service systems [40–42], but no studies have been conducted to analyze future investments in the trauma system network. In current practice, each component of the trauma network submits its requests to a central agency or decision-maker, where each request has a cost and a return. The central agency or decision-maker is responsible for allocating its limited budget among the requests. The cost of a request is in dollars, but the return is the improvement in patient outcomes. The impact of any set of investments on the system can be captured and estimated by simulation. In this paper, quantitative and computational methods for estimating these system investments will be examined and investigated.

Study Objectives. This study aims to develop a theoretical trauma system model and a general-purpose computational framework that facilitates the allocation and utilization of resources by the statewide trauma system.

The GTCNC desires to maintain and expand Georgia's trauma centers, strengthen emergency medical services in certain regions and develop an effective statewide transfer system [38]. To support the GTCNC's mission, we will model and replicate the Georgia Trauma Network and perform dynamic scenarios (that reflect the objectives of the GTCNC) to determine the trauma system funding allocation that would maximize the overall impact (e.g., patient outcome in terms of budgetary restraint, return on investment).

Given a local trauma network profile, injury distribution, and resource requests, the modeling and computational framework will enable (a) the creation of the set of all feasible and Pareto-efficient portfolios where limited funding is allocated across numerous resource requests from trauma centers, hospitals and EMS providers; (b) quantitative analyses of the impact of each feasible portfolio on the trauma system's performance measures via the Trauma System Simulator; and (c) sensitivity analyses to determine the best decision/policy to transport/transfer patients and to observe how possible changes in system inputs affect the return on investment and resource utilization.

The modeling framework enables decision/policy makers to analyze and understand the sensitivity of the system outcomes as a result of tactical and strategic decisions. The selection of investments that provides the best patient, hospital and EMS outcomes will benefit both the government finances and the public health.

2 Methods and Design

2.1 A Theoretical Model

Consider a trauma system network where there are T trauma centers (TC), H hospitals, N EMS providers, and M ambulances. Here "hospitals" denote those hospitals without a trauma center.

Assume there are Ti Level i TCs $(i = 1, ..., 4)$ and the statewide EMS system consists of R EMS regions, where each region covers a number of TCs, hospitals and EMS stations. The administrators at each of these sites submit upgrade requests to a central trauma system decision-maker. The costs of upgrade investments and their impact on the attributes of resources in the system are known. However, the exact financial benefit/return of each request is not known, and each differs in its impact on the system's performance. The central decision-maker must evaluate all the requests and select a portfolio of investments (the requests to be approved and to be refused) under a limited budget such that the selection will yield the best patient outcome.

Performance Measures. Measuring patient outcome is complicated and determining the best patient outcome requires definitive quantitative metrics that facilitate meaningful comparison.

We will investigate seven quantitative performance measures (Table 1) to explore the best representation of the trauma system in its efforts to improve patient care.

To improve outcome, it is desirable to reduce the number of patients waiting to be transported, reduce the time it takes a patient to get to a proper emergency care facility, reduce the ambulance travel time, increase the number of active ambulances, reduce the number of patients receiving lower quality care, and reduce the number of trauma

Table 1. List of seven performance measures

Metric	Description
M1	Average number of patients waiting to be transported to the hospital
M2	Average time between incident and patient arrival to TC (minutes)
M3	Average deadhead miles time per ambulance (minutes)
M4	Average number of ambulances in service
M5	Average number of patients in all hospitals
M6	Total number of patients who receive lower-quality care
M7	Proportion of trauma patients who received lower-level trauma care

patients unable to receive the proper level of trauma treatment based on their injury. These six metrics can be minimized or maximized as part of the system objectives. Metric 5 (M5) provides different insights by tracking the overall patient throughput based on the investment.

Type of Requests. The return of a portfolio can be measured by the percentage change in the performance metrics if that portfolio of investments is chosen. Since selection of one request is not independent of the selection of other requests, the value of returns for each portfolio will be different from each other. Table 2 shows a sample of requests, with associated costs and expected returns.

Table 2. Description of different type of requests submitted to the central trauma system decision-maker

Request	Type of request/description	Cost
Request Type I: Trauma center level upgrade		
Upgrade 1	Level II TC to Level I TC	$500k
Upgrade 2	Level III TC to Level II TC	$250k
Upgrade 3	Level IV TC to Level III TC	$150k
Request Type II: A preventive plan for a region		
Prevent 1	Reduce incidents by 11%	$90k
Request Type III: An equipment upgrade		
Equipment 1	Can treat patient 6% faster at that site	$50k
Equipment 2	Can treat patient 5% faster at that site	$50k
Request Type IV: Purchase of ambulance		
Ambulance 1	Add a heavy-duty ambulance	$260k

We will focus on four types of submissions:

- *Type I: trauma center level upgrade request* is the most expensive upgrade. A higher-level TC is typically better equipped to provide sufficient trauma care for the patients than a lower-level TC. However, the overall change in the system depends on which TC is upgraded and from which level and the number of trauma patients in each region. Naturally, TC upgrades may change pre-hospital patient flow. For example, some patients who are in serious condition and would have been transported to another TC can now be treated/transported to the upgraded TC. The change in patient flow may affect the arrival rate, utilization of other TCs, ambulance assignments and patient outcome throughout the system. Thus, measuring the change in the overall trauma system performance is not straightforward due to the interdependency of the system components.

- *Type II: preventive measure request* includes initiatives that are designed to reduce the frequency of trauma incidents. Some examples include preventing child maltreatment, preventing motor vehicle injuries, preventing falls among older adults, etc. [43]. According to the American College of Surgeon Committee on Trauma, trauma systems must develop prevention strategies to help control injury as part of an integrated, coordinated, and inclusive trauma system [44]. In our case, we classify different preventive plans into types according to the effect they have when adopted. It is also assumed that (from historic clinical and trauma data) the impact of a preventive plan in a region is known as the percentage reduction of injuries in that region.

- *Type III: Equipment upgrade* Some TCs may lack certain equipment and resources to provide the highest quality trauma care. Hence, a TC equipment upgrade can reduce patient treatment time and enable higher quality and safer treatments. Similar to previous types of requests, different equipment upgrades incur different costs and impact on the trauma system.

  We will consider equipment upgrades in terms of units, so a TC may request a unit of equipment upgrade funding. It is presumed that the impact of an equipment upgrade is known and given by the percentage reduction in patient treatment time (from clinical and trauma historic data). We note that equipment upgrades could impact other variables as well. We caution that speed is not the only variable impacting patient outcomes. For example, new equipment can improve diagnosis accuracy, resulting in better outcomes.

- *Type IV: Ambulance purchase* EMS providers may request the purchase of a new ambulance, particularly if they have a tough time satisfying incoming demand. In this case, expanding the fleet of ambulances is usually a workable solution, especially if it is proven to be cost-effective. If the request is approved, the region the EMS provider is responsible for will receive a new ambulance. While it should yield improvement in response times to patients, quantifying the amount of achieved improvement is implicit in the system.

Mathematical Formulation. Let $i = 1,..., I$ be the indices of the type of submissions and a_i be the number of submissions for each type. Given a_i for all i, the number of total submissions for the central decision maker is $A = \sum_{i=1}^{I} a_i$. Let $j = 1,..., A$ denote the indices for each submission. If $a_i > 1$, then at least two of the submissions are the same type. Finally, let c_j be the cost of submission j. Given a total budget B, the central decision-maker needs to allocate among the requests. Obviously, if the total cost of submissions $\sum_{j=1}^{A} c_j$ does not exceed B, then all the submissions can be approved. If not, we have to consider all the combinations where the total cost does not exceed B.

Let x_j be a binary variable to denote if submission j is selected or not, $j = 1, \ldots, A$. And let D be a $|R|x|A|$ adjacency matrix such that

$$D_{rj} = \begin{cases} 1 & \text{if submission } j \text{ is from EMS region } r \\ 0 & \text{otherwise} \end{cases}$$

Then the portfolio optimization to approve a submission can be formulated as:

$$(\textbf{TS-OPT}) \quad \text{Optimize} f(Q, x, m)$$

$$\text{s.t.} \sum_{j=1}^{A} c_j x_j \leq B \tag{1}$$

$$Dx \geq 1_{|R|} \tag{2}$$

$$Q(x) \geq m_{base} \tag{3}$$

$$x_j \in \{0, 1\} j = 1, \ldots, A$$

Here $1_{|R|}$ is a $|R|$-vector of all 1's. The objective function f is multi-objective, representing the expected improvement in performance (with respect to multiple measures, e.g., Table 1) from each selected request, as well as the interdependency and composite systems improvement Q with respect to the selected portfolio.

Equation (1) is the budget constraint. Equation (2) is the regional constraint reflecting that each EMS region should receive at least one submission approval. Equation (3) is the performance constraint to ensure the current system performance, m_{base}, will not be degraded after the investment.

Given the interdependency and cascading effect of the trauma system operations, it is analytically difficult to track and express in closed form the overall performance effects produced by a certain portfolio of requests. Specifically, since the trauma system is modeled as an integrated operation network of heterogeneous entities of trauma facilities, hospitals, and EMS providers, a change in one component affects various components (and is not necessarily linear). Thus, both functions f and Q in the (**TS-OPT**) formulation are not in closed forms.

2.2 An Integrated Computational Framework

Designing a simulation-optimization framework and a Trauma Simulator allows us to realistically model a working trauma system and its operations. It also serves as a useful tool to conduct quantitative analyses and comparisons on the trauma system characteristics and performance. The computational system facilitates a top-down approach in strategic resource allocation and decision-making: (a) the central decision-maker evaluates all the requests and establishes the set of all possible investment portfolios; (b) then, using the simulator, he/she analyzes and observes the performance effect of each portfolio to make the final informed decision. The computational platform also allows for the incorporation of uncertainty and experimentation with different parameters and distributions, which can be useful for change management and disaster emergency planning and training.

Below, we will outline a two-stage approach for the decision process. First, given a limited budget, a set of maximal feasible portfolios will be selected. Second, for each maximal feasible portfolio, a simulated trauma system with the corresponding resources and attributes will be generated to analyze and evaluate the overall system performance. The seven performance measures will serve as quantitative surrogates.

Stage 1: Finding the Pareto Feasible Investment Request Set. Suppose there are K different combinations of investments, which we call 'portfolios.' Let $k = 1,..., K$ denote the indices of all feasible portfolios. Let

$$y_{kj} = \begin{cases} 1 & \text{if submission } j \text{ of portfolio } k \text{ is approved} \\ 0 & \text{otherwise} \end{cases}$$

be the binary variable that specifies whether a request in a portfolio is approved or not.

Combinatorically, there exists an exponential number of portfolios in which the total cost is under the given budget. In practice, and for the sake of maximizing the return, decision-makers want to invest as much of their budget as possible; hence, we seek solutions that are maximal within the budget constraints, i.e., we search for the set of maximal feasible portfolios. This allows us to eliminate a large number of portfolios. The problem can now be described as finding all the portfolios that have a cost of at most B, with no portfolio having any remaining funds that can support a potential submission. The number of portfolios is further constrained by the requirement that there is at least one request from each region.

Mathematically, given A investments/submissions and their costs, and K portfolios, let $y_k \in \{0, 1\}^A$ denote a portfolio of investments $[y_{k1}, y_{k2},...,y_{kA}]$, $k = 1,...,K$; and $y_{kj} \in \{0, 1\}$ denote if submission j is selected in portfolio $k, j = 1, \ldots, A$. Let U_k be the set of submissions that are not selected in portfolio k. Our goal is to find all the non-dominated portfolios y_k subject to the following constraints:

$$\sum_{j=1}^{A} c_j y_{kj} \leq B \, \forall \, k \tag{4}$$

$$B - \sum_{j=1}^{A} c_j y_{kj} < \min_{j \in U_k} c_j \, \forall \, k \tag{5}$$

$$Dy_k \geq 1_{|R|} \, \forall \, k \tag{6}$$

Equation (4) models the budget constraint so that each portfolio selected is within the given budget. Equation (5) ensures that remaining funds from the budget of any portfolio must always be less than the minimum cost of the unselected submissions. Equation (6) reflects that each EMS region should receive at least one submission approval. Any portfolio that satisfies these conditions is called a maximal feasible portfolio and the set of such portfolios is called the maximal feasible portfolio set.

If there are A submissions, then the number of distinct portfolios is 2^A. We develop an efficient algorithm to generate all the maximal feasible portfolios.

We note that Stage 1 does not differentiate the impact of each submission. This is due to our discussion earlier: (a) for some submissions, such as upgrading the level of TCs, the impact of implementation is not known; and (b) the overall outcome obtained in the trauma system through the interaction of several submissions to be implemented is unknown. In other words, it is not possible to prioritize the investments since each investment yields improvement in different metrics of the trauma system. Furthermore, there is intrinsic uncertainty and interdependency in the overall patient impact. Hence, it is necessary to evaluate all the possible combinations that a feasible portfolio can take. We propose an algorithm that takes the costs of all submissions and the size of the budget as inputs and produces the maximal feasible portfolio set.

The following recursive algorithms (Algorithms 1, 2) perform the task of obtaining the feasible portfolio set.

Algorithm 1: Procedure to iterate all cases
Result: The set of maximal feasible portfolios
Initializations
for $i = 1$ to A **do**
| $base(i) \leftarrow 0$
end
$k = 0$
// Loop function $IterateOne(k,B,base)$ over k
while $k < A$ **do**
| $IterateOne(k,B,base)$
| $k++$
end

Algorithm 2: *IterateOne(cc,B,base)*
Input: Cost of each investment(*cost*) and size of budget (*B*)
Result: Feasible portfolios for one iteration are added to *the set of feasible portfolios*
Initializations
// Sort investments from smallest to largest cost
sortedCost ← sort(cost)
begin
| **if** *sortedCost*[*cc*] > B **then**
| | addToFeasibleInvestmentSet (base)
| **else**
| | remainingBudget ← B − *sortedCost*[*cc*]
| | *newbase* ← updateBase (base,cc)
| | **for** *j* = cc+1 to A **do**
| | | nextBudget ← remainingBudget
| | | - *sortedCost*[*j*]
| | | **if** remainingBudget ≥ 2 x *sortedCost*[*j*] **then**
| | | | IterateOne(j,remainingBudget,*newbase*)
| | | **end**
| | | **else if** (nextBudget < *sortedCost*[*j*] **and**
| | | (nextBudget < min(*sortedCost*)) **and**
| | | (nextBudget ≥ 0) **then**
| | | newbase2 ← updateBase(*newbase*)
| | | addToFeasibleInvestmentSet(*newbase2*)
| | | **end**
| | **end**
| **end**
end

Algorithm 1 iterates the subroutine *IterateOne()* in Algorithm 2 for the index set between *0* and *A*. This ensures that all possibilities for investments are covered, and that all feasible portfolios are added to the list.

In Algorithm 2, the cost array of all submissions, 'cost', is sorted from the smallest to the largest and stored in the array 'sortedCost'. Next, the array 'base' is defined as a vector of binary variables representing a portfolio; the array entries change in each iteration as the algorithm adds new submissions to the portfolio. So, starting from the array 'sortedCost', add investments to array 'base' until the total cost exceeds the budget limit, *B*. This must be done in a systematic way so that all feasible portfolios are generated without repetition.

Specifically, given an index number, Algorithm 2 checks whether the cost of submission at that index is greater than the remaining budget. If it is, we stop and add the 'base' array to the feasible investment set. If not, we will continue to investigate all other investments with a greater index independently. Add the investment on index j if cost[j] < remaining budget, where j > index. Also, the remaining budget is updated by subtracting the cost of the investment that was added last if it satisfies three main conditions:

- Condition (a) *If the updated remaining budget > cost of investment that is added last,* then continue to iterate the procedure with the given inputs (j, updated remaining budget, updated base array). This means if we decide to add an investment to the portfolio, we must continue to apply the same procedure starting from the index where we arrived last. In a recursive manner, we check all the possibilities for unarrived indices and add investments to the portfolio if there is budget available for that specific investment; and we then update the remaining amount of the budget.
- Condition (b) *If remaining budget < cost of investment that is added last AND the remaining budget > cost of the cheapest investment,* then do not add the investment at that index to avoid repetition of the same investment in the portfolio. It is apparent that the algorithm does not proceed to search for new investments if the remaining budget is less than the cost of the investment that is added last because the costs of investments with greater indices are greater than the remaining budget. So, at this point, the 'base' array must be the portfolio. However, it is a Pareto-dominated portfolio if the second condition, the remaining budget, is greater than the cost of the cheapest investment, is satisfied. The reason is since the array of costs have been sorted, the procedure will arrive at the cheapest options at some point in the procedure and if the remaining budget is greater than the cost of the cheapest investment, it means that the cheapest option has not yet been added to the portfolio. Also, if we add the cheapest investment to the portfolio, it will be the same as a portfolio that was created before with the recursive method in Condition (a). This condition always leads the algorithm to the correct path due to the sorting of costs from the smallest to the largest.
- Condition (c) *If remaining budget < cost of investment that is added last AND remaining budget < cost of the cheapest investment,* then stop. Do not add the next investment because there is insufficient remaining budget to add an investment to the portfolio. We then add the array 'base' to the set of feasible portfolios.

The feasible set of portfolios consists of arrays of 0–1 variables, where 1 indicates approval and 0 indicates refusal of submission at that index. For our problem, the inputs are summarized in Table 3. There are 89 submissions. Given the cost structure, the number of feasible portfolios is greater than two million. The effect of each portfolio on the trauma system will be evaluated using the in-house Trauma System Simulator as described below.

Stage 2: Designing the Trauma System Simulator. We assume that trauma incidents happen at random times at a rate P/year. This implies that each year, P trauma incidents occur that involve a patient to be transported to a trauma facility. The moment a trauma incident occurs, it is assumed that only one person is involved, and the location of the patient is generated by the steps described next. First, we determine the county where the incident occurred. The probability that an incident happens in a county is proportional to the ratio of injuries in one county to the injuries in the state. A county is modeled as a square with a center, and patient coordinates are generated uniformly in a square where its area is equal to the area of the county. Patient arrival is also determined by patient transfers from a lower-level TC or non-trauma hospital. In addition, it is assumed that patients' conditions may differ, and each patient is assigned an injury severity level according to the ratio of patients who have been treated in Level I, Level II, Level III, and level IV TCs in the past.

Recall there are N EMS providers and M ambulances belonging to the EMS providers, where $M > N$. The exact coordinates of EMS stations are known. It is assumed that ambulances are evenly distributed among the EMS stations. Ambulances are not only busy with trauma patients but also other types of patients, such as cardiac patients. Therefore, in the simulation at any point in time, ambulances are available with a certain probability.

The location and capacities of all TCs and hospitals are known. Capacity is defined as the number of trauma units in a TC and the number of trauma beds allocated in hospitals. We did not specifically consider the amount of human resources or special equipment to define the capacity because the usage of those resources is very complicated in the hospital environment. The topic of efficient management of resources in the hospital is out of the scope of this paper. A TC or a hospital must admit the patients who arrive. Once they are admitted, if there are any available trauma units, treatment of a patient starts. Patients stay in the hospital for a period of time, the length of which is determined by a random variable (and the severity of the generated trauma case), and they are either moved from the emergency department or transported to another facility to receive better treatment if necessary.

The simulation is intended to model the trauma system with their components, interactions, and decisions. The system works as follows: An incident occurs, a trauma patient is created, and then emergency services are called to transport the patient to a hospital. EMS assigns the closest available ambulance to the address where the call has been made. An ambulance crew responds to the assignment, and quickly drives to the address. Once they arrive at the incident point, they conduct the first intervention and transport the patient to the ambulance. The target hospital or TC is decided according to a procedure, which is summarized in Algorithm 3. The ambulance transports the patient to the most appropriate TC or hospital (based on the trauma incident level) and returns to its station once the patient is delivered to the hospital. A patient's treatment starts if trauma units or beds are available; if there is none, then the patient enters a priority queue, where he/she waits until the next trauma unit becomes available for treatment. In general, a trauma bed becomes available when the patient occupying that bed is moved to a non-trauma bed or is transported to another facility.

Algorithm 3: Procedure to decide patient's TC or hospital
Input: Patient's coordinates and location of all TCs and hospitals
Result: Decision of TC or hospital where the patient must be transported
Check all the TCs within r-mile radius from patient's location. Define set S as the
    set of trauma centers within the radius.
case *If there are any TCs with higher level in S* **do**
| Send the patient to the closest TC with higher or
| equal level.
case *If there are only equal AND lower level TCs in S* **do**
| Send the patient to the closest equal level TC
case *If there are only equal level OR lower level TCs in S* **do**
| Send the patient to the closest TC
case *If no TC within the radius. S* **do**
| Send the patient to the closest hospital since there is no TC within the radius.
end

If the patient was transported to a lower-level trauma center or an ordinary hospital, this patient is a candidate for transfer to a higher-level TC to receive better and/or more suitable treatment for his/her injury. With a certain probability, the patient's severity of injury may reduce, and there is no need to transfer the patient to another TC center. However, if the injury remains too severe and requires the patient to be treated at another TC center, then the process of patient transfer is initiated. The patient transfer process is almost identical to the process of responding to a first-time patient. (Note that in a rural/semi-rural hospital, a helicopter is often used because of distance and traffic.)

In a transfer situation, the location of the patient is the hospital where he/she is treated. Similarly, the EMS is called, and the closest available ambulance is assigned for patient transfer; when the ambulance arrives at the hospital, the patient is picked up and transported to the most appropriate TC. There is a decision-making process to determine which trauma center the patient should be transferred to. In this case, we ignore whether there is any TC within a radius r; instead, we simply send the patient to the closest available TC, where the level of TC is greater than or equal to the patient's injury severity. If there is no available higher-level TC due to capacity limitations, then the patient is sent to the closest available TC, where the TC level is higher than the current facility's level. If there is none in this case as well, which is quite exceptional, the patient is left in the current hospital. Treatment continues for a certain time and then the possibility of patient transfer is considered again. Figure 1 shows the trauma patient flow schema within the trauma system.

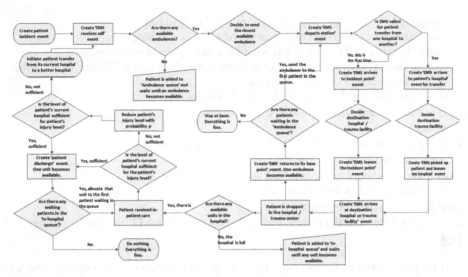

Fig. 1. An abbreviated flowchart to illustrate how the trauma system works.

2.3 Data, Experimental Design and Outcome Measures

Data Collection and Processing. For the simulation, we developed a theoretical layout of the Georgia trauma system. Arrival rates of patients, coordinates of TCs, hospitals, EMS stations, injury related statistics, number of ambulances, hospital and TC capacities, and population and regional statistics of counties of Georgia are found from various public resources, including the Centers for Disease Control and Prevention, the GTCNC and the Georgia Association of Emergency Medical Services. The relevant parameters used in the simulation model are given in Table 3.

Table 3. Value of Parameters Used in the Model

Parameter	Value	Unit
Patient arrival rate per year (P) (Increase by increments of 5,000)	25,000	Patients
Budget to allocate for investment requests (B)	5 million	$
Parameters about trauma centers, hospitals, and ambulances		
Number of Level I TC's (T_1)	5	TC
Number of Level II TC's (T_2)	9	TC
Number of Level III TC's (T_3)	6	TC
Number of Level IV TC's (T_4)	5	TC
Number of non-trauma hospitals (H)	110	Hospital

(*continued*)

Table 3. (*continued*)

Parameter	Value	Unit
Number of EMS stations (N)	285	Station
Number of ambulances (M)	2300	Ambulance
Radius of circle to scan TC's to decide destination TC (r)	30	Miles
Probabilistic parameters		
Probability that an ambulance is busy with other type of patients at any time	90%	
Probability that patient type I's injury severity level reduces to level II	5%	
Probability that patient type II's injury severity level reduces to level III	10%	
Probability that patient type III's injury severity level reduces to level IV	15%	
Process times distributions		
Duration between incident and first call to EMS (call time)	~ Normal(3.5, 1)	Minutes
Duration between call time and ambulance departure from hospital (preparation time)	~ Normal(3, 1)	Minutes
Duration to carry the patient to ambulance at emergency scene (carry time)	~ Normal(5, 1)	Minutes
Duration of patient stays in the hospital (treatment time)	~ Normal(480, 60)	Minutes
Number of submissions by type		
Level II TC - > Level I TC	9	
Level III TC- > Level II TC	6	
Level IV TC- > Level III TC	5	
Reduce incidents by 11%	10	
Can treat patient 6% faster at that site	50	
Add ambulance	9	

The exact incident time of patients are generated following a Poisson distribution because this type of data could not be accessed due to confidentiality issues. Process times throughout the simulations are assumed to follow a normal distribution with certain parameters, where negative random variables are omitted. In addition, the number of submissions by type has been given as well. The TC ambulance requests are submitted by the TCs, EMS stations and the EMS regions. It is not important which element of the system submits the request; it is the methodological framework that is being developed where one can apply data from anywhere to it to perform an analysis that matters.

Outcome Measures. The outcome measures can be categorized into three classes:

a) *Patient related statistic:* This includes the number of patients transported in a year, the number of patients transferred from one facility to another facility, the average time between the incident and the patient's arrival to the destinated TC, average waiting time for a trauma unit of a patient in the hospital etc.

b) *EMS related statistics:* This consists of the average deadhead mile time per ambulance, the average response times per ambulance and average time that an ambulance is active.

c) *TC or hospital related statistics:* This includes the average number of patients waiting in the queue of the trauma unit, the proportion of patients who have waited in the queue to the total number of patients who have arrived and average utilization.

Note that some measures may be an implied component of one or more classes. While patient outcome is an important measure, it is difficult to quantify objectively.

Three metrics among the outcome measures, one from each of these three classes, are chosen to assess the overall performance of the trauma system. The chosen metrics are (a) average time between incident and patient's arrival to the destinated TC, (b) average deadhead mile time per ambulance and (c) proportion of patients who have waited in the queue to the total number of patients who have arrived. The first metric is patient-related due to the importance of transporting the patient to a TC or hospital as soon as possible. The second metric is ambulance-related since average deadhead miles give EMS an idea about the ambulance utilization and is often used in computing the financials of EMS. The last metric is TC-related, since patients who have to wait in the queue do not receive sufficient quality and timely care in the TC, and the proportion of that number to the total number of patients arrived to the TCs gives an idea about the quality of the trauma care and treatment in that TC.

The basic results that are obtained via the Trauma System Simulator consist of these three metrics. Comparisons among portfolios will be made with respect to the three metrics, since they have been considered good indicators of the overall performance in the system.

Experimental Analysis. First, we are interested in understanding the interplay and tradeoffs among the system performance measures when demand changes. Initially, the trauma incident rate is set to 25,000 patients per year. We will analyze how the system responds when patient arrival rate surges in increment of 5,000 to 50,000 patients per year. These values are selected based on the estimated trauma incidents per year in Georgia (covering the years 2016–2021). Georgia's registry reported 36,192 trauma incidents in 2020 [45].

Second, we are curious about the scanning radius of ambulances and how it may affect patient flow between scenes of emergency to the TCs. A quick sensitivity analysis reveals that a smaller radius is always better in all scenarios.

Moreover, there is a very strong, positive correlation between the magnitude of radius for scanning and the values of all performance measures. This implies that when transporting trauma patients from incident locations to a TC/hospital, EMS personnel should consider only the closest TC/hospitals as the destination candidates. Including hospitals further from the incident location may result in the selection of TC/hospitals farther away, thus increasing the transportation time and from our sensitivity analyses, degrading all other performance measures.

Census data and geographic locations of all trauma centers in the U.S. reveal that approximately 85% of the U.S. population (both urban and rural) is within thirty miles of a trauma center. Given these results and how similar Georgia's specific proportion of Urban-to-Rural populations is to the U.S. proportions, for our analysis, we use thirty miles as our ambulance search radius from the trauma incident location.

3 Results

3.1 Generating the Pareto-Efficient Frontier

For each maximal feasible portfolio obtained in Stage 1, simulation is run using the Trauma System Simulator for a period of two years. The simulation is repeated 40 times using different seeds of patient information, including incident location, injury severity level and incident time, to ensure that the results are statistically consistent. Table 4 shows some sample results for the incident rate of *35,000* per year and Table 5 reports the associated best portfolios list. Each reported portfolio is the best with respect to one or more metrics.

Table 4. Best observed values of outcome metrics by seeds (a sample only)

Seed #	Average time between incident and patient arrival to TC (minutes)	Average deadhead mile time per ambulance (minutes)	Proportion of trauma patients who received lower level of trauma care
1	54.498624	35.25873	0.003879
2	54.897654	35.246306	0.003634
3	55.560384	35.519721	0.004540
4	55.605276	35.437128	0.004423
....			
39	55.498146	35.438552	0.004566
40	55.581468	35.525422	0.004623
Avg	55.240506	35.385691	0.004196

It is observed that for each metric, numerous portfolios achieve the best results, and they could be different among the seeds. We record which portfolios are marked as the best portfolio for each metric and count the number of times they were observed among the 40 seeds. The best results are tabulated in Table 6. However, counting does not inform us that some portfolios are significantly dominant over others. Therefore, we need to define a procedure (Algorithm 4) such that we can decide on a portfolio eventually.

We caution that since this is a three-objective simulation-optimization problem, these results are Pareto-optimal, as there does not exist a portfolio in which all three metrics achieve the best values simultaneously.

Table 5. Index of best portfolios for each metric in each seed

Seed #	Average time between incident and patient arrival to TC (minutes)	Average deadhead mile time per ambulance (minutes)	Proportion of trauma patients who received lower level of trauma care
1	160	160	359
2	2602	2919	2603
3	4002	4002	3598
4	4541	4596	4589
....			
39	13074	13074	13262
40	13558	13558	13960

Table 6. Values of outcome metrics obtained in the final selection of investments.

Set #	Average time between incident and patient arrival to TC (minutes)	Average deadhead mile time per ambulance (minutes)	Proportion of trauma patients who received lower level of trauma care
160	54.8403	35.2539	0.00236
7152	55.2292	35.4406	0.00161
9268	54.8752	35.2692	0.00253
11458	55.5022	35.5353	0.00168

3.2 Selecting a Restricted Set of Top-Ranked, Pareto-Efficient Portfolios

Each portfolio (a point) is plotted on a 3-D graph, with each axis associated with one metric. The convex hull of all the points estimates the feasible manifold of the best portfolios, when evaluated under the three metrics.

Here, the vertices of the convex hull are of interest. The set of vertices are examined in 2-D for each pair of metrics, and for each step, the Pareto efficient portfolios are marked. This reduces the number of candidate portfolios to a smaller number. Once Pareto efficient portfolios are found for all pairwise comparisons, the procedure aims to find a portfolio where it is observed in the maximum number of sets of pairwise comparisons. If there exists only one such portfolio, it is decided as the best portfolio. If multiple portfolios exist, then the procedure focuses on common investments approved among portfolios. It finally reaches a conclusion, where it may be either one unique portfolio or a set of portfolios. The decision to select the final portfolio is left to the decision-maker. A formal description of the procedure is as follows:

Algorithm 4
Let

$P = \{25,000, 30,000, 35,000, 40,000, 45,000, 50,000\}$, the set of trauma incidents per year

$M = \{1, \ldots, 3\}$, the set of performance metrics

$K = \{1, 2, \ldots, 2,304,028\}$, the set of portfolio ids

$S = \{1, 2, \ldots, 40\}$, the set of seeds

Let X_{psmk} denote the value of metric m for portfolio k, obtained by running simulation using the sth seed with trauma incident rate p.

We index the 3 chosen outcome metrics as $m = 1, 2, 3$, and define $d_{psm} = \{k : \min_i \{X_{psmk}\}\} \forall p \in P, \forall s \in S, m = 1, 2, 3$ as the ID of the portfolio where the minimum value of metric m is attained when the simulator is run with parameter p and seed s. Without loss of generality, the vector d_{ps} is represented by $[d_{ps1}\ d_{ps2} d_{ps3}]^T$.

We use a 3×3 matrix, z_{ps}, to store the values of each performance measure observed for the portfolio in which a minimum is attained for one of the outcome metrics. In matrix form z_{ps} is given by

$$z_{ps} = \begin{bmatrix} X_{ps1d_{ps1}} & X_{ps2d_{ps1}} & X_{ps3d_{ps1}} \\ X_{ps1d_{ps2}} & X_{ps2d_{ps2}} & X_{ps3d_{ps2}} \\ X_{ps1d_{ps3}} & X_{ps2d_{ps3}} & X_{ps3d_{ps3}} \end{bmatrix}$$

Let w_p be a $(|S| \times |M|) \times (|M|+1)$ matrix where $(z_{ps} d_{ps})$, $s \in S = \{1, \ldots, 40\}$, $M = \{1, 2, 3\}$ are appended. In our case, the matrix w_p has 120 rows and 4 columns. Specifically,

$$w_{ps} = \begin{bmatrix} \begin{pmatrix} z_{p1} & d_{p1} \\ \vdots & \vdots \\ z_{p40} & d_{p40} \end{pmatrix} \end{bmatrix}$$

The first three entries of each row of matrix w_p can be viewed as a point in R^3, and there is a total of 120 3-D points. The convex hull of these points gives a set of points and a corresponding portfolio index. Let

$$C_p = \left\{ c: \text{point indicated in } c^{th} \text{row of } w_p \text{ is on the convex hull} \right\}$$

be the set of indices on the convex hull of points represented by w_p.

What follows is to obtain the Pareto-efficient frontier for all pairwise comparison of three metrics (dimensions). Since there are three metrics, it comes with $\binom{3}{2} = 3$ possible 2-D spaces, and all the points on the Pareto-efficient frontier are recorded.

Let $E_p = E_{p(1-2)} \bigcup E_{p(1-3)} \bigcup E_{p(2-3)}$ be the union of elements on the Pareto-efficient frontiers for all pair of dimensions. Then, we define

$$EI_p = \left\{ h : \max_{h \in E_p} \left\{ 1_{\{h \in E_{p(1,2)}\}} + 1_{\{h \in E_{p(1,3)}\}} + 1_{\{h \in E_{p(2,3)}\}} \right\} \right\} \forall p$$

as the set of indices h, where any $h\prime \neq h \in E_p$ is observed in all the sets $E_{p(1-2)}, E_{p(1-3)}, E_{p(2-3)}$ no more than $h \in E_p$. Here we try to find portfolio(s) that is (are) the most common among the 2-D efficient frontier index sets. Once EI_ps are found for $\forall p \in P$, it is possible to track the process backwards to find the best portfolios.

We then look at the set of portfolio ID's corresponding to the indices that have been most observed in all Pareto-efficient sets of two dimensions. If the set size is one, it implies there is a unique best portfolio for the given parameter p. If the set size is greater than one, it indicates that there are multiple portfolios where they are non-dominated with respect to their impact on performance measures.

3.3 Incorporating Costs

After completing a systematic reduction of the maximal feasible portfolios, we apply cost to further reduce the remaining portfolios.

Let

$$OF_p = \left\{ f : \min_{f \in EI_p} \left\{ cost \; of \; portfolio \; f \; \in EI_p \right\} \right\}$$

be the set of portfolios in which minimum cost is achieved in the set EI_p. It is noteworthy to emphasize that $s\,(OF_p) > 1$ is possible for any $p \in P$. In this case, the final decision is left to the priorities of the decision-maker.

Application of this procedure to our problem produced the selected portfolios in Table 6 with their corresponding performance measures. The selection of investments has not been displayed simply due to the sheer size of the investment set.

Comparing the four selected portfolios in Table 6, we observed that

a) They all request a reduction of trauma incidents through prevention methods.
b) They all request two units of equipment upgrade (total to $100,000).
c) All except one portfolio (i.d. #160) requests at least one ambulance.
d) Portfolio 160 includes an upgrade of a Level IV TC to Level III TC

3.4 Interdependency of Performance Measures

We investigate the interplay and dependency of the seven quantitative performance measures (Table 1) to explore the best representation of the trauma system in its efforts to improve patient care.

Empirical results show that Metrics 1 and 2 have a strong positive correlation and they behave in a parallel manner with respect to all the other metrics. These two metrics align perfectly since long waiting queue of trauma patients at the scene translates to longer incident to TC/hospital time, and vice visa. Similarly, a positive correlation exists between Metrics 3 and 4. Metrics 1 and 2 have a strong positive correlation with Metric 3 and Metric 4, as seen in Fig. 2. This translates to the longer it takes an ambulance to respond to a call and get the patient to the TC, the more patients will end up waiting at incident locations. These values are also related to the average number of miles an ambulance must travel to pick up the patient and to get to the TC. The longer the distance, the longer the response and patient delivery take. All of these are again related to the number of ambulances in service.

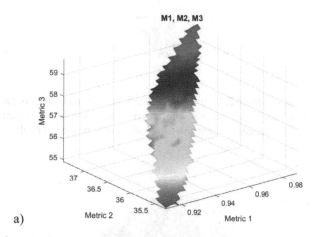

a)

Fig. 2. The surface plot for Metric 1, Metric 2 and Metric 3 shows the strong positive correlation of Metric 1 and 2 with Metric 3. The longer the deadhead mile time for ambulances, the longer it takes for the response; hence, there is a longer time between the incident site to the TC/hospital, and the longer the patient queue at the scene.

There is a slight positive correlation between Metrics 1 and 4 and Metric 5 and Metric 7. Metrics 1 and 4 appear to have no correlation with Metric 6. Metric 6 and Metric 7 also have a very strong positive correlation, but that is simply because Metric 7 is calculated using Metric 6. When Metric 6 is observed with the other metrics as shown in Fig. 3, we can easily observe that local minimums and maximums spread throughout the linearly increasing trend line. There is a slight positive correlation between Metrics 1 and 4 with Metric 6, but due to the spikes, we see that the number of patients receiving low-quality care is not always consistent with higher average times.

Metric 5 does not appear to have any correlation with Metric 6 or Metric 7 as shown in Fig. 4, so the average number of patients in hospitals is not related to the number of TC patients receiving a lower-level care. The average number of people in all hospitals maintains a very tight range. Overall, this means that regardless of the number of patients, hospitals stay busy at the same rate, which is dependent on the number of beds. This also reflects the fact that trauma patients have higher priority over non-trauma patients in bed access.

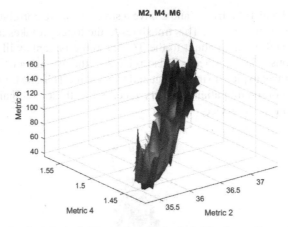

Fig. 3. A surface plot for Metric 2, Metric 4, and Metric 6. The fluctuations of valleys and peaks indicate that the total number of patients who receive lower-quality care (Metric 6) does not necessarily correspond to longer incident site-to-hospitals time or more ambulances in service.

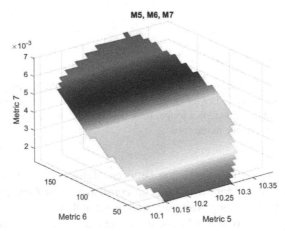

Fig. 4. A surface plot for Metric 5, Metric 6, and Metric 7. The average number of patients in hospitals (Metric 5) does not influence the number of TC patients receiving a lower-level care (Metric 7), reflecting that trauma patients have higher priority in bed access. Metric 7 is derived from Metric 6; hence they correlate to each other.

4   Limitations

Modeling the entire trauma system network is complicated. We aim to produce a system that can realistically reflect the on-the-ground situation without the need for perfect data sources. Some limitations include:

Inter-incident Time for Trauma Incidents.   The availability of data on trauma incidents requires a comprehensive and well-maintained database of records of trauma incidents. Not every state establishes a comprehensive trauma registry; moreover, ethics, patient privacy and data protection law must be adhered to. Although hospitals, emergency medical services and trauma centers may maintain records of trauma incidents, accessing such data requires cooperation and permission from all relevant healthcare institutions. While we do have access to several major Level 1 and Level II trauma centers, we opt to assume that the inter-incident times follow a Poisson distribution with rate P patients per year instead of generating estimates using the partial data that we have access to.

Trauma Bed Capacity. Trauma bed capacity reflects the number of available hospital beds that are specifically designated and equipped for the treatment and care of trauma patients. These beds are often located in specialized units, such as trauma centers, emergency departments, or intensive care units (ICUs), and are equipped with the necessary resources and personnel to provide immediate and comprehensive care to patients with severe injuries.

We can obtain the trauma bed capacity for all but one trauma center. In the one missing site, a minimum number of trauma beds is assumed. Since non-trauma care hospitals do not have designated trauma beds, it is assumed that their trauma patient capacity is similar to the average trauma patient capacity of trauma centers (which are known).

We remark that the trauma bed capacity in a hospital or healthcare facility depends on several factors: hospital type and size, patient volume, hospital resources, regional demands, and flexible capacity. It is important to note that trauma bed capacity (even for a trauma center) is dynamic and may change based on the hospital's ability to adapt to patient needs.

Assignment of Trauma Level to Patients. The assignment of trauma level to each patient is complicated and multi-factorial. In the model, we assign trauma level according to the injury severity score of the patients; but the decision of the destination hospital does not solely depend on this. However, the complexity of these types of details during decision-making is simplified in our model.

Specifically, assigning a trauma level to patients typically involves a process known as "triage." Triage is the systematic process of quickly assessing and prioritizing patients based on the severity of their injuries or medical conditions. The goal of triage is to ensure that the most critically injured or ill patients receive immediate care while also addressing the needs of less severe cases. The general steps for assigning trauma levels to patients in a triage setting include scene safety, immediate care, patient assessment, documentation, and reassessment.

It is also important to note that the specific triage categories and procedures can vary depending on the healthcare facility, the level of training of the personnel and the nature of the incident.

Number of Ambulances at Each Station. The number of ambulances stationed at a particular location or station can vary widely based on several factors, including the size of the service area, the population it serves, the level of need and the budget of the EMS provider. We could not find the number of ambulances for each station; rather than using partial knowledge, we assume that the ambulances are evenly distributed among the stations based on regional population density and historical call volumes.

In general, the specific number of ambulances at each station is determined based on the careful consideration of numerous factors (e.g., size of service area, population density, call volume, response time goals, types of emergencies, shift scheduling, budget and resources, specialized units and overall patient transport needs) to ensure an effective and timely response to emergencies. EMS providers often conduct data analysis and planning to allocate resources optimally, which are based on the needs of the community and their available resources. The decision often reflects a balance between providing adequate coverage and managing costs efficiently.

5 Conclusion and Future Research

Trauma systems are vital components of healthcare infrastructure, and they play a crucial role in improving patient outcomes and reducing mortality and disability resulting from traumatic injuries. They also play a key role in disaster preparedness, helping hospitals and healthcare providers plan for and respond to mass casualty incidents, ensuring that resources are available when needed most.

Trauma systems are designed to ensure a rapid and coordinated response to traumatic injuries, with trauma centers within these systems staffed with highly trained and specialized medical professionals, including trauma surgeons, anesthesiologists, and critical care nurses. This expertise is crucial for managing the complexities of traumatic injuries. The National Study on the Costs and Outcomes of Trauma identified a 25% reduction in mortality for severely injured patients who received care at a Level I trauma center rather than at a non-trauma center. Triage protocols are used to assess and prioritize the care of patients to ensure that the most critically injured patients receive immediate attention, optimizing their chances of survival and recovery.

By designating specific facilities as trauma centers and distributing trauma care resources strategically, trauma systems decision-makers are tasked to ensure that resources are allocated efficiently. This improves the care of trauma patients and contributes to overall community health and safety.

In this study, we design a dynamic mathematical model and computational framework to facilitate an effective resource allocation of trauma funds across the State of Georgia. The results presented here are subject to change when applied to a different state or regional trauma system, as there would be different parameters, cost structures and submissions. However, the most important contribution of this study is that it establishes a general-purpose computational framework of investment allocation for trauma systems and ways to model, optimize, and evaluate the impact of the investments on the overall performance of the trauma system.

Given a local trauma network profile, injury distribution and resource requests, the modeling and computational framework enables the creation of the set of all feasible and Pareto-efficient portfolios where limited available funding is allocated across numerous requests (investments) from trauma centers, hospitals, and EMS providers. Using the framework, decision-makers can quantitatively analyze the impact of each feasible portfolio on the trauma system's performance measures via the Trauma System Simulator. Sensitivity analysis can then be conducted to determine the best decision or policy to transport or transfer patients and to observe how changes in system inputs affect the return on investment and resource utilization.

The design is a top-down approach at a strategic level which simultaneously uses tactical level decisions to evaluate several strategies to improve the system. Systems simulation-optimization is a powerful interlacing tool to perform a thorough analysis and systematic upgrade of a trauma system with given investments, which facilitates the decision-making process of trauma commission leaders.

The procedure described in the Results Section reduces an overly enormous number of candidate portfolios to a reasonable number by exploiting the existence of Pareto-efficient frontiers of the values obtained for specified metrics. This type of approach can be applied to a broad class of problems involving the selection of investments and resource allocations by considering their impact on certain performance measures. The procedure can be extended by: (a) increasing the number of performance measures (i.e., increasing the dimension of the problem), (b) assigning a certain utility function dependent on the performance metrics for the patient and performing the elimination process with the functions, or (c) increasing the time span of interest in the problem and adding a dynamical perspective to the selection of best portfolios.

While some of our recommendations have been implemented by Georgia's trauma commission leaders, it is important to realize that the empirical experiments conducted are intended as a proof-of-concept to demonstrate that improvements to patient well-being are an independent outcome and can be measured in some manner and that alternative solutions can be analyzed. Results from the simulation runs are easily interpretable; thus, the findings can support the decision-makers in their choice of investments by supplying the expected outcomes.

We caution that appropriate inputs should be used to conduct analyses that are relevant to the specific population (arrival rates of patients; locations of trauma centers, hospitals, and EMS stations; injury and time-related statistics; number of ambulances; hospital and trauma center capacities; and population and regional statistics of counties). The validity and strength of these intuitive arguments can be assessed via the computational performance of the Trauma System Simulator using different settings and parameters.

In a follow-up study, we will derive and analyze a risk-driven, multi-objective portfolio optimization model [46] that will offer the incorporation of uncertainties in the theoretical model (**TS-OPT**) formulated in Sect. 2. Future research should involve the application of this modeling framework using real world inputs for the portfolio options and the investment level. This can produce potentially more interesting results due to the reality of limited funding and the distinct types of investment most often available. Other research should investigate other scenarios, such as the difference in portfolios dedicated to urban trauma system upgrades versus rural trauma system upgrades. Lastly, a significant area of interest is pediatric trauma care due to the severely constrained resources available to this patient group. Dedicating a trauma simulation model to measuring how to improve pediatric trauma care could be very impactful.

Acknowledgement. This work is partially supported by grants from the National Science Foundation IIP-0832390 and IIP-1361532 and the Georgia Trauma Care Network Commission. Findings and conclusions in this paper are those of the authors and do not necessarily reflect the views of the National Science Foundation or the Georgia Trauma Care Network Commission.

References

1. Nathens, A.B., Brunet, F.P., Maier, R.V.: Development of trauma systems and effect on outcomes after injury. Lancet **363**(9423), 1794–1801 (2004). https://doi.org/10.1016/S0140-673 6(04)16307-1
2. MacKenzie, E.J., Rivara, F.P., Jurkovich, G.J., et al.: The national study on costs and outcomes of trauma. J. Trauma Acute Care Surg. **63**(6), S54–S67 (2007). https://doi.org/10.1097/TA. 0b013e31815acb09
3. Mullins, R.J., Veum-Stone, J., Hedges, J.R., et al.: Influence of a statewide trauma system on location of hospitalization and outcome of injured patients. J. Trauma Acute Care Surg. **40**(4), 536–546 (1996). https://doi.org/10.1097/00005373-199604000-00004
4. Hulka, F., Mullins, R.J., Mann, N.C., et al.: Influence of a statewide trauma system on pediatric hospitalization and outcome. J. Trauma Acute Care Surg. **42**(3), 514–519 (1997). https://doi. org/10.1097/00005373-199703000-00020
5. Candefjord, S., Asker, L., Caragounis, E.C.: Mortality of trauma patients treated at trauma centers compared to non-trauma centers in Sweden: a retrospective study. Eur. J. Trauma Emerg. Surg. **48**(1), 1–12 (2022). https://doi.org/10.1007/s00068-020-01446-6
6. Danner, O.K., Wilson, K.L., Heron, S., et al.: Benefit of a tiered-trauma activation system to triage dead-on-arrival patients. W. J. Emerg. Med. **13**(3), 225 (2012). https://doi.org/10.5811/ westjem.2012.3.11781
7. Haas, B., Stukel, T.A., Gomez, D., et al.: The mortality benefit of direct trauma center transport in a regional trauma system: a population-based analysis. J. Trauma Acute Care Surg. **72**(6), 1510–1517 (2012). https://doi.org/10.1097/TA.0b013e318252510a
8. Ariss, A.B., Bachir, R., El Sayed, M.: Factors associated with survival in adult patients with traumatic arrest: a retrospective cohort study from US trauma centers. BMC Emerg. Med. **21**(1), 1–10 (2021). https://doi.org/10.1186/s12873-021-00473-9
9. Rittenberger, J.C., Callaway, C.W.: Transport of patients after out-of-hospital cardiac arrest: closest facility or most appropriate facility? Ann. Emerg. Med. **54**(2), 256–257 (2009). https:// doi.org/10.1016/j.annemergmed.2009.01.009
10. Taylor, B.N., Rasnake, N., McNutt, K., McKnight, C.L., Daley, B.J.: Rapid ground transport of trauma patients: a moderate distance from trauma center improves survival. J. Surg. Res. **232**, 318–324 (2018). https://doi.org/10.1016/j.jss.2018.06.055
11. Cameron, P.A., Zalstein, S.: Transport of the critically ill. Med. J. Aust. 169(11–12) (1998). https://doi.org/10.5694/j.1326-5377.1998.tb123434.x
12. Schneider, A.M., Ewing, J.A., Cull, J.D.: Helicopter transport of trauma patients improves survival irrespective of transport time. Am. Surg. **87**(4), 538–542 (2021). https://doi.org/10. 1177/0003134820943564
13. Han, H.R., Miller, H.N., Nkimbeng, M., et al.: Trauma informed interventions: a systematic review. PLoS ONE **16**(6), e0252747 (2021). https://doi.org/10.1371/journal.pone.0252747
14. Richardson, M., Eagle, T.B., Waters, S.F.: A systematic review of trauma intervention adaptations for indigenous caregivers and children: insights and implications for reciprocal collaboration. Psychol. Trauma **14**(6), 972–982 (2022). https://doi.org/10.1037/tra0001225
15. Hamilton, S.M., Breakey, P.: Fluid resuscitation of the trauma patient: how much is enough? Can. J. Surg. **39**(1), 11 (1996)
16. Haukoos, J.S., Byyny, R.L., Erickson, C., et al.: Validation and refinement of a rule to predict emergency intervention in adult trauma patients. Ann. Emerg. Med. **58**(2), 164–171 (2011). https://doi.org/10.1016/j.annemergmed.2011.02.027
17. Awwad, K., Ng, Y.G., Lee, K., Lim, P.Y., Rawajbeh, B.: Advanced trauma life support/advanced trauma care for nurses: a systematic review concerning the knowledge and skills of emergency nurse related to trauma triage in a community. Int. Emerg. Nurs. **56**, 100994 (2021). https://doi.org/10.1016/j.ienj.2021.100994

18. Stojek, L., Bieler, D., Neubert, A., Ahnert, T., Imach, S.: The potential of point-of-care diagnostics to optimise prehospital trauma triage: a systematic review of literature. Eur. J. Trauma Emerg. Surg. **49**(4), 1–13 (2023). https://doi.org/10.1007/s00068-023-02226-8

19. Vles, W.J., Steyerberg, E.W., Meeuwis, J.D., Leenen, L.P.H.: Pre-hospital trauma care: a proposal for more efficient evaluation. Injury **35**(8), 725–733 (2004). https://doi.org/10.1016/j.injury.2003.09.006

20. Savoia, P., Jayanthi, S., Chammas, M.: Focused assessment with sonography for trauma (FAST). J. Med. Ultrasound **31**(2), 101–106 (2023). https://doi.org/10.4103/jmu.jmu_12_23

21. Kannikeswaran, N., Ehrman, R.R., Vitale, L., et al.: Comparison of trauma and burn evaluations in a pediatric emergency department during pre, early and late COVID-19 pandemic. J. Pediatr. Surg. **58**(9), 1803–1808 (2023). https://doi.org/10.1016/j.jpedsurg.2023.03.008

22. Amato, S., Vogt, A., Sarathy, A., et al.: Frequency and predictors of trauma transfer futility to a rural level I trauma center. J. Surg. Res. **279**, 1–7 (2022). https://doi.org/10.1016/j.jss.2022.05.013

23. Veenema, K.R., Rodewald, L.E.: Stabilization of rural multiple-trauma patients at level III Emergency departments before transfer to a level I regional trauma center. Ann. Emerg. Med. **25**(2), 175–181 (1995). https://doi.org/10.1016/S0196-0644(95)70320-9

24. Halvachizadeh, S., Teuber, H., Cinelli, P., Allemann, F., Pape, H.C., Neuhaus, V.: Does the time of day in orthopedic trauma surgery affect mortality and complication rates? Patient Saf. Surg. **13**(1), 1–8 (2019). https://doi.org/10.1186/s13037-019-0186-4

25. Prin, M., Li, G.: Complications and in-hospital mortality in trauma patients treated in intensive care units in the United States, 2013. Inj. Epidemiol. **3**(1), 1–10 (2016). https://doi.org/10.1186/s40621-016-0084-5

26. Tote, D., Tote, S., Gupta, D., Mahakalkar, C.: Pattern of trauma in a rural hospital and factors affecting mortality in trauma patients. Int. J. Res. Med. Sci. **4**(2), 450–456 (2016). https://doi.org/10.18203/2320-6012.ijrms20160294

27. Atlas, A., Büyükfirat, E., Ethemoğlu, K.B., Karahan, M. A., Altay, N.: Factors affecting mortality in trauma patients hospitalized in the intensive care unit. J. Surg. Med. **4**(11), 930–933 (2020). https://doi.org/10.28982/josam.812409

28. Sammy, I., Lecky, F., Sutton, A., Leaviss, J., O'Cathain, A.: Factors affecting mortality in older trauma patients - Aasystematic review and meta-analysis. Injury **47**(6), 1170–1183 (2016). https://doi.org/10.1016/j.injury.2016.02.027

29. Hefny, A.F., Idris, K., Eid, H.O., Abu-Zidan, F.M.: Factors affecting mortality of critical care trauma patients. Afr. Health Sci. **13**(3), 731–735 (2013). https://doi.org/10.4314/ahs.v13i3.30

30. Palmer, C.: Major trauma and the injury severity score - where should we set the bar? In: Annual Proceedings - Association for the Advancement of Automotive Medicine (2007)

31. Cobianchi, L., Dal Mas, F., Agnoletti, V., et al.: Time for a paradigm shift in shared decision-making in trauma and emergency surgery? Results from an international survey. World J. Emerg. Surg. **18**(1), 14 (2023). https://doi.org/10.1186/s13017-022-00464-6

32. Keene, C.M., Kong, V.Y., Clarke, D.L., Brysiewicz, P.: The effect of the quality of vital sign recording on clinical decision making in a regional acute care trauma ward. Chinese J. Traumatol. English Ed. **20**(5), 283–287 (2017). https://doi.org/10.1016/j.cjtee.2016.11.008

33. Little, W.K.: Golden hour or golden opportunity: early management of pediatric trauma. Clin. Pediatr. Emerg. Med. **11**(1), 4–9 (2010). https://doi.org/10.1016/j.cpem.2009.12.005

34. Rogers, F.B., Rittenhouse, K.J., Gross, B.W.: The golden hour in trauma: dogma or medical folklore? Injury **46**(4), 525–527 (2015). https://doi.org/10.1016/j.injury.2014.08.043

35. Brathwaite, C.E.M., Rosko, M., McDowell, R., Gallagher, J., Proenca, J., Spott, M.A.: A critical analysis of on-scene helicopter transport on survival in a statewide trauma system. J. Trauma Injury, Infect. Crit. Care **45**(1), 140–146 (1998). https://doi.org/10.1097/00005373-199807000-00029

36. Duchin, E.R., Neisinger, L., Reed, M.J., Gause, E., Sharninghausen, J., Pham, T.: Perspectives on recovery from older adult trauma survivors living in rural areas. Trauma Surg. Acute Care Open. **7**(1), e000881 (2022). https://doi.org/10.1136/tsaco-2021-000881

37. Flynn, D., Francis, R., Robalino, S., et al.: A review of enhanced paramedic roles during and after hospital handover of stroke, myocardial infarction and trauma patients. BMC Emerg. Med. **17**(1), 1–13 (2017). https://doi.org/10.1186/s12873-017-0118-5

38. Georgia Trauma Commission. Georgia Trauma System Strategic Plan – Five Years (2021). https://trauma.georgia.gov/planning

39. MacKenzie, E.J., Hoyt, D.B., Sacra, J.C., et al.: National inventory of hospital trauma centers. JAMA **289**(12), 1515–1522 (2003). https://doi.org/10.1001/jama.289.12.1515

40. Wu, C.H., Hwang, K.P.: Using a discrete-event simulation to balance ambulance availability and demand in static deployment systems. Acad. Emerg. Med. **16**(12), 1359–1366 (2009). https://doi.org/10.1111/j.1553-2712.2009.00583.x

41. Aboueljinane, L., Sahin, E., Jemai, Z.: A review on simulation models applied to emergency medical service operations. Comput. Ind. Eng. **66**(4), 734–750 (2013). https://doi.org/10.1016/j.cie.2013.09.017

42. Lubicz, M., Mielczarek, B.: Simulation modelling of emergency medical services. Eur. J. Oper. Res. **29**(2), 178–185 (1987). https://doi.org/10.1016/0377-2217(87)90107-X

43. Centers for Disease Control and Prevention. *Injury Prevention and Control.* http://www.cdc.gov/injury/

44. Trauma System Evaluation and Planning Committee. Regional Trauma Systems: Optimal Elements, Integration and Assessment (2008). https://www.facs.org/media/sgue1q5x/regionaltraumasystems.pdf

45. Georgia Trauma Annual Report. 2021. Georgia Department of Public Health. https://dph.georgia.gov/trauma

46. Leonard, T., Lee, E.K.: US-Mexico border: building a smarter wall through strategic security. J. Strateg. Innovation Sustain. **15**(1), 156–182 (2020). https://doi.org/10.33423/jsis.v15i1.2735

Dispersion of Personal Spaces

Jaroslav Horáček[1]([⊠]) [iD] and Miroslav Rada[2,3] [iD]

[1] Faculty of Humanities, Department of Sociology, Charles University,
Prague, Czech Republic
jaroslav.horacek@fhs.cuni.cz
[2] Faculty of Informatics and Statistics, Department of Econometrics,
Prague University of Economics and Business, Prague, Czech Republic
[3] Faculty of Finance and Accounting, Department of Financial Accounting
and Auditing, Prague University of Economics and Business, Prague, Czech Republic
miroslav.rada@vse.cz

Abstract. There are many entities that disseminate in the physical space
– information, gossip, mood, innovation etc. Personal spaces are also entities that disperse and interplay. In this work we study the emergence
of configurations formed by participants when choosing a place to sit
in a rectangular auditorium. Based on experimental questionnaire data
we design several models and assess their relevancy to a real time-lapse
footage of lecture hall being filled up. The main focus is to compare the
evolution of entropy of occupied seat configurations in time. Even though
the process of choosing a seat is complex and could depend on various
properties of participants or environment, some of the developed models
can capture at least basic essence of the real processes. After introducing
the problem of seat selection and related results in close research areas,
we introduce preliminary collected data and build models of seat selection
based on them. We compare the resulting models to the real observational
data and discuss areas of future research directions.

Keywords: Seat preference · Personal space · Territoriality ·
Computer modeling

1 Introduction

Most of us have encountered a situation when we entered a lecture room or
auditorium for the first time and had to choose a seat. A few people had been
already seated there and we expected an indefinite number of people to arrive
after us.

It is very interesting to observe formation of such patterns from the perspective of a teacher or speaker. For example, at the beginning of his book [8]
Schelling contemplates a phenomenon that occurred during his lecture – several
front rows of seats were left vacant whilst the other rows in an auditorium were
fully packed.

Despite its common occurrence, phenomena corresponding to patterns resulting from seating preferences are not frequently studied. There is only few studies

H. Moosaei et al. (Eds.): DIS 2023, LNCS 14321, pp. 159–169, 2024.
https://doi.org/10.1007/978-3-031-50320-7_11

dedicated to this topic. In [2] they study the influence of seating on study performance. The study [5] shows that students are biased to choose seats on the left side of the classroom. In [4] they review the impact of seating arrangement.

An important factor influencing the seat selection process is territoriality. The study [6] identified seats that are considered highly territorial – seats at the end of rows or in the corners of a classroom. It was showed that highly territorial seats attract students with greater territorial needs. Various arrangements of an auditorium were examined – auditorium consisting of rows of tables with chairs, U-shaped auditorium, rows of tablet-arm chairs or chairs clustered around isolated tables.

In another study [3] they consider interplay of human territoriality and seating preference. They discovered that after several lectures students tend to occupy the same seat (or at least location). More than 90% of students confirmed that they tend to always occupy the same location. However, they were unsure about the reasons for such a behavior. The following reasons received the highest scores – security, sense of control or identity. However, there are several other reasons for this behavior such as social order, social role, stimulation etc.

Human territoriality might have several meanings unlike the territoriality of animals [3]. One of the possible meanings could be privacy. It is useful to understand it as control of access to other people. In [7] they use Privacy Preference Scale (PPS) that is based on this definition. It distinguishes various types of privacy – intimacy, not neighboring, seclusion, solitude, anonymity, reserve. Particularly seclusion and solitude need some further explanation. Seclusion is understood as being far from others, whilst solitude means being alone with others nearby. All variants of privacy might be related to seating preference. In this study, rows in a lecture hall were marked as front, middle and back. Then various privacy scores of participants seated in those regions were measured. Most of the privacy factors were more important for participants seated towards the back. Nevertheless, some of the privacy conditions could be reached independently of seating in the back or front, e.g., seclusion, anonymity or not neighboring.

There are other studies focused on seating outside the lecture hall. Choosing a seat in a crowded coffee-house is studied in [9]. In his thesis [10] Thomas focuses on various situations in public transport. One section is devoted to spaces between passengers. There are documented examples of maximization of distances among passengers. If necessary, people tend to sit next to people expressing similarity (e.g., similar gender) or passengers not expressing a defensive behavior (e.g., placing a body part or other item on adjacent seat). The average distance of two passengers was measured to be 1.63 seat units. Another work apply agent-based models to study emergence of clusters in a bus [1].

Most of the mentioned studies are descriptive – they try to statically capture the properties of seating arrangements or participant deliberation about seat selection. The purpose of this work is to design dynamic models of seat selection that can shed new light on the formation process of collective seating patterns. In the next section we describe what factors do we consider when building such models. In the third and fourth section we describe the preliminary experimental data and the setup for comparing models respectively. The next section is devoted to designing the models and their comparison. In the final section we summarize the results and outline future research directions.

2   Factors to Consider

Let us imagine that a person (or a group of people) arrives in an auditorium and seeks a suitable place to sit. We will call such a person *participant*.

There are several factors that could influence participant's seat preference. One of them could be visual or olfactory attractiveness of already seated participants. We were often asked about these assumptions during experimental phase of this work. They could play a significant role, however we assume that all participants are basically neutral in appearance. We are interested in situations when there is no such disruptive element in the auditorium.

There are several types of influence outside the environment. It makes a difference if we know the speaker or whether we expect to be asked questions during a lecture. That could have a significant influence on our seat selection. Hence we assume that there is no familiarity or interaction during a lecture.

Another factor we encountered in our experiments is the length of a lecture. Some of the participants would choose to sit at the ends of seat rows. Either because of having an escape way if the lecture is not satisfactory or to have easy access to rest rooms.

Two other factors are mentioned in [8]. First, people might have learned the seating habits from other activities or events. For example, they have had a bad experience with sitting in the front. The second is a hypothesis that people just do not think about the seat selection too much. They just take any near vacant seat, that satisfies some basic social rules, e.g. not too many people need to stand up when getting to a seat. The book suggest that it could be due to the of interest. However, we suggest that the reason could be the limited time of deliberation on seat selection. We return to this idea when designing seat selection models.

In the introductory section the two important factors were mentions – territoriality and privacy. We will definitely make use of them in the following sections.

Seating preferences could be based on sensory capabilities of a participant. For example, a participant with hearing or vision impairment would rather sit in the front. For now, we do not include these conditions in our models.

In [1] they consider the following parameters of participants – age, gender, size, ethnicity. We do not take these parameters into account either since results on relation of these parameters and seating preferences are not apparent. In [6] they observed a higher territoriality with women, however, it could be explained simply by the fact that observed woman participants carried more items and hence needed more personal space.

Another characteristics more common for public transport is shifting seats as mentioned in [10]. If ones seat is not satisfactory it is reasonable to change it. In real situations we can observe people switching their seats with somebody else or moving to a different location. Such a behavior is not included in our models but it is a subject of future work.

3   Data and Its Evaluation

To obtain the first results we rely on preliminary empirical data. As the first source we used publicly available time lapse videos of lecture halls being filled up. We were able to find several videos, however, one represented the process of seating selection in a medium sized lecture hall particularly well[1].

In the footage, participants arrive alone or in small groups, hence we also consider adding new participants in bunches. This footage serves as the first inspiration for further research, because it neatly illustrates the formation of seating patterns.

We followed the video until we were still able to track the order of seating of the participants. At that moment the majority of seats still remained unoccupied. Nevertheless, such a moment, when the auditorium is still empty enough, poses the most interesting case. When the auditorium is nearly full, then the seating preference is overpowered by the need of a seat.

After extracting the visible part of the auditorium, there remained 7 rows of seats. Each row contained 14 seats. Therefore, the auditorium can be represented by a rectangular grid of size 7×14, where each cell represents one seat. In Fig. 1 we show such a representation. The black cells represent occupied seats. The numbers in black cells denote the order of arrival of corresponding participants. If more cells share the same number it means that all these participants arrived at the same time as a group.

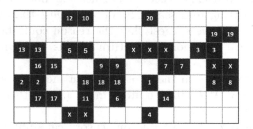

Fig. 1. A tabular depiction of the real auditorium filled with participants. The black cells represent occupied seats. The white cells represent empty seats. The seats marked with X represent participants seated before the start of recording. The number in the black cell depicts the order of arrival of the corresponding participant.

To understand the process of seating pattern formation we took the same approach as in the paper [9]. We prepared and printed four schematic depictions of seating arrangements in a 7×14 auditorium with some already seated

---

[1] At the time of writing the paper (14 October, 2023), the footage remains publicly accessible at the link https://www.youtube.com/watch?v=r76dclwZU9M.

participants. It was represented in the tabular way mentioned above. The examples were based on the initial seating arrangements from the real footage. The preliminary data collection took two rounds. Several undergraduate university students were asked whether they want to participate in the following short thought experiment. All respondents participate in ICT study programs. The collection of data was voluntary and anonymous. We did not collect any personal information such as age, gender etc., since there is no proven influence of these factors on seating preference.

In the first round, 12 students were asked about 4 seating situations depicted in Fig. 1. They were asked to mark their preferred seating position. If several positions seemed equivalent, they were asked to mark all of them. The first round served as a pretest of quality of the example seating configurations and we used it also to improve the overall experiment setup.

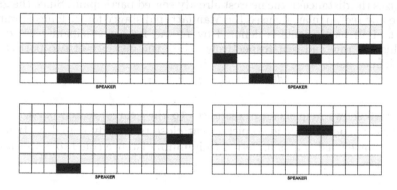

Fig. 2. Four initial example configurations. The position of a speaker is marked for each hypothetical auditorium tableau.

In the second round there participated 27 students (not participating in the first round). The students were presented the same four hypothetical situations as in the first round, since they proved to be useful. This time, they we asked to select only one position of their preferred seat. They were instructed as follows: *"Imagine you being a participant at a lecture. It is a one-time event. You participate voluntarily and are interested in the content of the lecture. You arrive from the back of the auditorium and need to pick one seat. Both sides of the auditorium are well accessible. You do not know anybody of the participants. The lecture is not long (maximum duration is 1 h). There is no expectation of you being asked questions by a speaker during the lecture. Some people are already seated in the auditorium. They all seem neutral in appearance. You should definitely expect some more people to come. However, you do not know the amount of the newcomers. Which is your most preferred seat?"*

Out of the 27 respondents, 9 marked exactly the same spot for all four configurations. There are several possible reasons for this outcome. First, they were bored with the task and hence filled the form without paying attention. Second, the seat choice depended on a preferred seat position in general and was hence independent of the seating arrangement (this was verbally confirmed by one of the participants). Anyway, in this preliminary testing phase we temporarily leave the 9 respondents out of our analysis since we do not know anything about their motivation fur such answers. Interestingly, 8 of these 9 respondents chose a position at the edge of the auditorium.

For our further analysis we work with 18 valid questionnaires. Since each consisted of 4 configurations, it gives us 72 cases of marked seating preferences in total.

There are several interesting outcomes of the analysis. First, if there is enough space, respondents rarely choose a position directly neighboring (vertically, horizontally or diagonally) to someone. Such a choice was made by only 3 participants in only 6 cases (about 8% of all the cases).

We first worked with the hypothesis that participants choose a seat that maximizes the distance to the nearest already seated participant. Since the grid is discrete and rectangular, we used the Manhattan distance for such measurement. The set of all participants is denoted by P and for a participant p we denote p_r, p_s his/her row and seat respectively. Then the distance of two participants p, q is defined as:

$$d(p, q) = |p_r - q_r| + |p_s - q_s|.$$

Figure 3 shows frequencies of distances to the nearest seated participant. It seems that participants tend to pick a seat close to already seated participants. However, some small gap is preferred (minimum distance equal to 3 appeared in 43% of cases).

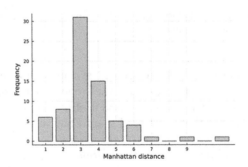

Fig. 3. Frequency of distances to the nearest seated participant.

Contrary to our hypothesis, participants do not seem to maximize their minimum distance to others. Only in 4 out of 72 cases could we see a seating choice that visually resembled maximization of distance to others. In the discarded set of participants a seat selection similar to maximization is more common (12 out of 36 cases). Nevertheless, maximization of distance to other participants is a phenomenon we return to in the next sections.

Another way to measure distance is the distance from the center of the mass of already seated participants. We assign numbers to the rows starting from the back (the largest number is of the front row). Similarly, we number the seats from left to right. Then, row c_r and seat c_s of the center of mass c is defined as:

$$c_r = \text{round}\left(\frac{\sum_{p \in P} p_r}{|P|}\right), \quad c_s = \text{round}\left(\frac{\sum_{p \in P} p_s}{|P|}\right).$$

We use rounding to the nearest integer with round half up tie-breaking rule.

In Fig. 4 we show the frequency of distances to the center of the mass of already seated participants. For measurement we omitted the lower right configuration from Fig. 2 because there was only one seated groups. The largest distances were mostly produced by one respondent. It seems that a reasonable distance from the center of mass plays (maybe subconsciously) a role in the choice of a seat. We will utilize the results in these following sections.

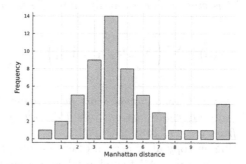

Fig. 4. Frequency of distances to the center of mass.

4   Basic Setup

Using the results from the previous section, it is our goal to design models that capture the dynamics of seating formation. The data obtained from the real footage serves for comparison of developed models.

For such a purpose the data from Fig. 1 were slightly modified. The cells where participants sit next to each other and have different numbers are marked as one group of participants. From the footage it is clear that when a participant sat directly next to another one, they were acquainted and basically had intended to sit next to each other from the beginning.

The general seating process is defined as follows. In one time step a group of participants arrives and chooses a space where all group members can sit next to each other. In the real world participants could leave empty seats among them in the case when the auditorium is not fully packed. Since the participants do not know the number of participants arriving after them, we do not consider such a possibility and we always perceive a group as one unit that needs to be placed compactly.

Every simulation run shares the same initial configuration of already seated participants that is defined by "x" marks in Fig. 1. We use the same size of the auditorium (7 × 14). That means that a group of 2, 2 and 3 participants respectively is already seated in the auditorium at the prescribed positions. Next, we add groups of exactly the same sizes and in the same order as in the real footage. We always assume that there is at least one spot where a group can fit.

At every time step a configuration of already seated participants forms a certain pattern. To capture the essence of entropy of such a pattern we devised the following simple measure. For every row it counts the number of transitions from an empty seat to a seated person and vice versa. For every row this count is squared and then these values are summed up. This is done to emphasize the rows of auditorium. This way an arrangement of densely packed seated participants will obtain a small score whereas participants loosely spread across the auditorium with plenty of free space among them will get significantly higher score. We can represent an auditorium as a binary matrix A where $A_{i,j} = 1$ if the j-th seat in the row i is occupied. The matrix element $A_{i,j}$ is equal to 0 when the corresponding seat is empty. For an auditorium of size $m \times n$ the entropy measure is calculated as

$$e(A) = \sum_{r=1}^{m} \left(\sum_{s=2}^{n} |A_{r,s-1} - A_{r,s}| \right)^2 .$$

Example: *Just one row with 14 occupied seats has entropy 0. One row where empty and occupied seats alternate has entropy 169. The entropy of the final seating arrangement from the real footage is 231.*

By measuring seating entropy after arrival of each group we can capture the evolution of the entropy in time. All tested models are stochastic – they use randomness to choose among equivalently good seating positions. Therefore, for each model of seat selection the evolution of entropy in time is obtained by averaging 1,000 independent runs of the model.

5   Models of Seat Selection and Their Comparison

Using the analysis of the empirical data we design several models of seating behavior and compare them with the real data obtained from the video footage. In the previous section we pointed out the focus of such a comparison – the evolution of seating entropy in time.

In this section we present the models in the order of their creation. For each model we introduce a keyword that is used to address the model. The evolution trajectory of the real world data is marked as `real`. All models work in a greedy manner. An arriving group chooses a place to sit and remains seated there. The models are following:

- *Random seat selection* (`random`) – This model tries to implement the case when participants actually do not care about choosing a position and rather select an arbitrary empty seat. Hence, a group selects randomly a compact bunch of empty seats in one row where it can fit.

- *Maximization of personal space* (`max`) – A group simply chooses a seating that maximizes the minimum Manhattan distance to other seated participants. Even though it does not seem that majority of participants select a seat in this manner this model is implemented for the sake of contrast and comparison. Measuring the minimum distance for a group of participants is equivalent to measuring the distance for each group participant separately and then taking minimum distance.
- *Personal space selection* (`space`) – This model loosely incorporates the findings from the experiment data. A group tries to find an empty space where the minimum Manhattan distance to others is somewhere between 2 and 4 (including the bounds). If there is no such spot it tries to find a place at even larger distance. If such seat cannot be found, the group selects a random available space.
- *Simple space selection* (`simple`) – The following model is a combination of the first model and a simplification of the previous model. In this setup a group does not care about seating too much. However, it rather chooses some space where it has at least some amount of personal space, i.e., basically one empty seat is left around the group. This is operationalized by the group having a minimum Manhattan distance to all participants greater than 2. If there is no such spot then a seating is chosen randomly.
- *Center of mass selection* (`center`) – A group considers all available spaces that have minimum Manhattan distance to others greater or equal to 2. From such available positions it selects randomly among the ones that are closest to the center of the mass of already seated participants. If there is no such space available, then the group chooses its spot randomly.

In Fig. 5 we show comparison of the models. The last method (`center`) incorporating the information about the center of mass seems to resemble the actual real entropy evolution in the closest way. The model with simple minimization of distance to the center of mass seems The model is the only one that follows the "bump" of the real data in between the time slots 7 and 9.

Random selection model (`random`) and maximization model (`max`) provide the lowest resemblance to the real data. This supports the idea that participants neither do not select their seating positions randomly nor do they maximize distance from others. The simple variant with embedding the idea of personal space (`simple`) seems to be much more accurate. The most complex method (`space`) is outperformed by the simpler method (`simple`). Such an observation might be consistent with Occam's razor principle.

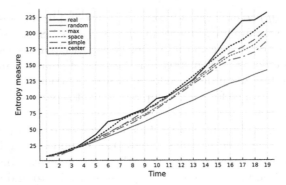

Fig. 5. Evolution of entropy in time for each of the predefined models and the real data. For each model the trajectory is averaged over 1,000 independent runs of the model.

6   Summary

Several entities could disseminate in the physical space – information, moods, memes, gossips. If we adhere to the previous studies of human territoriality and privacy we could understand personal space to be an entity that also disseminates in the physical space. Personal spaces of people intervene and force each other to form various patterns. Some of the patterns can be observed during the process of seating arrangements. Hence, the main focus of the work was studying the process of emergence of seating configurations in rectangular auditorium. Especially, when participants arrive in smaller groups.

In this work we utilized a time lapse footage of a lecture hall being filled up and questionnaires with seating examples in order to be able to form models of emerging seating patterns. The model where seat selection is based not only on reasonable distance to others but also on distance to the center of the mass of already seated participants seems to be the most plausible. Its evolution of auditorium entropy in time resembles most closely the entropy of data obtained from the real footage.

However, the data we used have limitations. The video footage is only one-time example of such process. Therefore, it does not capture the vast possibilities of seating arrangements. The questionnaire might suffer from being of artificial nature and not reflecting the true basis of seat selection process entirely.

In the real seat selection process many factors can play a role. We discussed them in the second section of this paper. Some of them could be neglected by limiting the modeled situations, some of them cannot be properly modeled (e.g., the attraction among participants), some of them need to be further employed in future models.

In future work we intend to capture our own footage of emerging seating arrangements. The method of using questionnaires also has its advantages, hence we intend to obtain much larger data collection with more refined test configurations. Especially, it is worth testing the current hypothesis about the influence of the center of the mass of already seated participants. Also some other test configurations are needed to enlighten the behavior of participants when an auditorium is nearly full. It would also be beneficial to enable a verbal feedback of participants for individual configurations.

Another interesting research direction is exploration of various measures of distance and entropy. In this work we mostly used the Manhattan distance measure and our simple custom entropy measure. Exploration of other variants might bring even more realistic approach to how participants perceive seating arrangements and also better model comparison.

One particular research question attracts us. A real seating arrangement is formed locally by people gradually coming to an auditorium. The final seating arrangement is obtained by a process we could call "crowd social computing". We can understand participants and their collectivity as a computational unit that is able by its own gradual self-organization to reach a certain configuration. Our question is how close (or how far) such a configuration is from a patterns computed globally in advance, i.e., by means of optimization. That is because several optimization areas deal with similar tasks, e.g., rectangle packing or, if we perceive the auditorium as a set of individual rows, bin packing.

Last but not least, results of studying the seat selection process can be applied to design of lecture or concert halls. Especially, to enable better use of their space and to help participants become more comfortable and concentrated.

Acknowledgements. The work of the first author was supported by Charles University, project GA UK No. 234322. The work of the second author was supported by the Czech Science Foundation project 23-07270S.

References

1. Alam, S.J., Werth, B.: Studying emergence of clusters in a bus passengers seating preference model. Transp. Res. Part C Emerg. Technol. **16**(5), 593–614 (2008)
2. Bergtold, J.S., Yeager, E.A., Griffin, T.W.: Spatial dynamics in the classroom: does seating choice matter? PLoS ONE **14**(12), e0226953 (2019)
3. Guyot, G.W., Byrd, G.R., Caudle, R.: Classroom seating: an expression of situational territoriality in humans. Small Group Behav. **11**(1), 120–128 (1980)
4. Haghighi, M.M., Jusan, M.M.: Exploring students behavior on seating arrangements in learning environment: a review. Procedia Soc. Behav. Sci. **36**, 287–294 (2012)
5. Harms, V.L., Poon, L.J., Smith, A.K., Elias, L.J.: Take your seats: leftward asymmetry in classroom seating choice. Front. Hum. Neurosci. **9**, 457 (2015)
6. Kaya, N., Burgess, B.: Territoriality: seat preferences in different types of classroom arrangements. Environ. Behav. **39**(6), 859–876 (2007)
7. Pedersen, D.M.: Privacy preferences and classroom seat selection. Soc. Behav. Pers. Int. J. **22**(4), 393–398 (1994)
8. Schelling, T.C.: Micromotives and Macrobehavior. WW Norton & Company, New York (2006)
9. Staats, H., Groot, P.: Seat choice in a crowded cafe: effects of eye contact, distance, and anchoring. Front. Psychol. **10**, 331 (2019)
10. Thomas, J.A.P.K.: The social environment of public transport. Ph.D. thesis, Victoria University of Wellington (2009)

An Evolutionary Approach to Automated Class-Specific Data Augmentation for Image Classification

Silviu Tudor Marc(✉)(iD), Roman Belavkin(iD), David Windridge(iD),
and Xiaohong Gao(iD)

Department of Computer Science, Middlesex University, London NW4 4BT, UK
SM2947@live.mdx.ac.uk

Abstract. Convolutional neural networks (CNNs) can achieve remarkable performance in many computer vision tasks (e.g. classification, detection and segmentation of images). However, the lack of labelled data can significantly hinder their generalization capabilities and limit the scope of their applications. Synthetic data augmentation (DA) is commonly used to address this issue, but uniformly applying global transformations can result in suboptimal performance when certain changes are more relevant to specific classes. The success of DA can be improved by adopting class-specific data transformations. However, this leads to an exponential increase in the number of combinations of image transformations. Finding an optimal combination is challenging due to a large number of possible transformations (e.g. some augmentation libraries offering up to sixty default transformations) and the training times of CNNs required to evaluate each combination. Here, we present an evolutionary approach using a genetic algorithm (GA) to search for an optimal combination of class-specific transformations subject to a feasible time constraint. Our study demonstrates a GA finding augmentation strategies that are significantly superior to those chosen randomly. We discuss and highlight the benefits of using class-specific data augmentation, how our evolutionary approach can automate the search for optimal DA strategies, and how it can be improved.

Keywords: Data Augmentation · Genetic Algorithm ·
Regularization · Hyperparameter Optimization

1 Introduction

As a field of study, computer vision strives to equip computers with the ability to accurately comprehend and efficiently process visual data, such as images and videos [6, 27]. Computer vision involves several sub-domains, such as scene reconstruction, object detection and recognition, and image restoration. Convolutional neural networks (CNNs), a specialized variant of deep feedforward networks, have gained widespread popularity as a powerful technique in computer vision

H. Moosaei et al. (Eds.): DIS 2023, LNCS 14321, pp. 170–185, 2024.
https://doi.org/10.1007/978-3-031-50320-7_12

[12]. These networks have demonstrated remarkable potential for achieving high generalization accuracy in classification tasks, as evidenced by their success in various applications such as high-resolution remote sensing scene classification [15], fashion item classification [11], facial recognition [16] and numerous other use cases. Among these, CNNs are used in the field of computer-aided diagnoses (CAD), such as liver lesion classification [10] and detecting lung diseases in X-ray images [3]. A CNN learns to identify patterns and objects in the data using a labelled dataset. This process involves iteratively updating the weights in the network so that the model can learn to classify images or data samples into different categories accurately. This training process aims to develop a CNN model that can generalize well to new data samples that it has not seen before. The capability of a CNN to apply its learned knowledge from labelled datasets and make predictions on new, unseen data is called generalization.

It is well-known that the generalization ability of a CNN is in direct relationship with the size and quality of the dataset. In this case, as the size and quality of the dataset used to train a CNN increase, the network's generalisation ability also increases in a directly proportional manner [20]. However, many applications have difficulties with the data collection process, as sometimes data is scarce or expensive to label [29]. A surrogate for data collection is data augmentation (DA), which increases the dataset synthetically. DA facilitates creating an artificially expanded training set by generating modified data from existing data points [29]. Two main types of DA commonly used in computer vision are global DA and class-specific DA. The former involves applying the same transformations to all images in the dataset, regardless of their class. Several studies have researched and proposed global DA strategies that improved the robustness and generalization ability of deep feedforward networks [24]. Although this can be beneficial in some cases, it can also result in suboptimal performance due to the differential class relevance of different transformations [13].

On the other hand, class-specific DA applies data transformations designed explicitly for each class in the dataset. The transformations used for each class may vary depending on the specific characteristics of the class, such as object orientation or lighting conditions [26]. Class-specific approaches to DA may be beneficial, which has been empirically demonstrated in several cases [13, 25] and [21]. DA can be considered an iterative process because it involves testing different augmentation techniques (e.g. rotation, contrast) and parameters (e.g. angles, limit) to see how they affect the performance of a machine-learning (ML) model. This process typically involves trial and error, as different datasets and models may require different augmentation methods and parameters to achieve optimal performance [23]. Various augmentation libraries, including Albumentations [4], provide an extensive selection of image transformation techniques, with over sixty options available. The availability of such a vast range of techniques further underscores the iterative nature of DA, as it may be necessary to evaluate multiple methods before arriving at an optimal approach for a given task. Additionally, implementing class-specific DA becomes more complex as the number of potential combinations increases exponentially with each additional class in the dataset. This exponential growth in possible combinations highlights the challenges of developing effective DA strategies for multi-class classification problems. The search space of class-specific DA can be computed as α^l, where α is the number of image transformations in a library, and l is the number of classes in the dataset. As a result, we are confronted with a potentially intractable combinatorial problem over a finite set of transformations. Being intrinsically combinatorial, no immediate gradient descent procedure is available, so the problem consequently cannot be absorbed into the underlying ML problem, being thus one of hyperparametric optimization. For this reason, metaheuristics were selected as our preferred approach. Metaheuristics represent a class of optimization algorithms adept at finding approximate solutions to complex optimization problems [18]. These algorithms are designed to explore large search spaces and identify high-quality solutions efficiently. Noteworthy examples of metaheuristic algorithms include simulated annealing, genetic algorithm (GA), ant colony optimization, particle swarm optimization, and tabu search. The diverse metaheuristic approaches make them popular for addressing optimization problems in various domains.

GA's are an evolutionary algorithm that takes inspiration from genetic principles, particularly natural selection. A GA utilizes various operators, including selection, mutation, and recombination, to mimic the evolutionary process and ultimately find an optimal solution to a given problem [2]. The application of evolutionary algorithms for global DA has been investigated in prior research such as [28] and [23]. Concurrently, various techniques for class-specific DA without employing any search algorithm have been suggested and have exhibited advancements in the capacity for generalization, such as those presented in [25] and [21].

To the best of our knowledge, there has been no prior research on using GA to address class-specific DA problems. Consequently, we introduce an automated search approach for class-specific DA in a classification task, utilizing a GA. The proposed framework comprises two algorithmic components: a GA and a CNN. The GA will be utilized to search for class-specific DA strategies doubling the dataset, while the CNN will be trained to extract features from the augmented dataset. The GA and CNN will work together in a fitness score-based testing approach, where the CNN will test the solutions found by the GA and provide a fitness score based on the final model generalization ability.

We present the results of our proposed framework on three datasets from the e-commerce and healthcare industries. The results demonstrate the ability of the GA to find better class-specific augmentation strategies with an improvement of up to 62.57% and up to 36.70% on the fashion and medical dataset, respectively, than the manually set global augmentation strategy. The results provide strong evidence of the effectiveness of our framework in improving the performance of CNN in image classification tasks.

The following structure is adhered to in the subsequent sections of this article. Initially, we provide an overview of various frameworks associated with class-specific DA and evolutionary techniques for automating the DA process. Subsequently, we introduce our methodologies and datasets. We then describe the experimental setups and metrics employed for evaluation. Lastly, we present the performance of our approach in the results section.

The code can be found at
https://anonymous.4open.science/r/EvoStarComparison-32E1/.

2   Related Works

Previous studies have proposed evolutionary-based global data augmentation (DA) frameworks. For example, Terauchi et al. [28] applied a thermodynamical genetic algorithm (GA) to an existing global auto-augmentation method (AutoAugment [5]), and proposed a solution to controlling the diversity of genotypes in the population. Their findings showed comparable improvement in accuracy while significantly reducing the search space and time of the search. Another example is presented in [23], where an evolutionary-based automatic DA search tool is developed to explore optimal global augmentation strategies. Despite the automation of the DA process, their findings indicate that the implemented strategies did not exhibit statistically superior performance compared to the manual strategies.

In addition, class-specific DA frameworks have been proposed. For instance, in [25], semantic attacks are used to generate new data in the case of object detection, where a small but imperceptible perturbation is introduced to the real-world data to cause the model to predict incorrectly. The model is then trained on the samples generated by the semantic attacks, resulting in improved average precision of a specific class and the overall map of the object detection network. Another study demonstrated that class-specific DA could improve ML performance on datasets with limited data, [21]. This paper's remaining sections will present our methods, datasets, experimental setups, and results.

3   Methods

In deep learning (DL), it is widely acknowledged that deep feedforward neural networks require a substantial amount of labelled data to achieve high levels of accuracy. As a result, the generalization performance of a convolutional neural network (CNN) is correlated with the size of the dataset [20]. Nevertheless, acquiring new real-life data is infeasible, arduous, or cost-prohibitive in numerous cases. For instance, the medical field may encounter such issues when dealing with rare diseases where data is scarce. Another example, companies may face obstacles when attempting to leverage DL models due to the expense or difficulty of generating or labelling datasets [9]. Data augmentation (DA) is a cost-effective approach that enables the synthetic enlargement of a dataset by manipulating the existing data. DA typically involves the following steps:

1. Selecting the augmentation methods: These may include rotation, translation, scaling, flipping, colour shifting, etc.

2. Applying the augmentation methods: This involves using libraries to automatically generate new examples by applying the selected transformations to the original images.
3. Adding the augmented data to the dataset: A more extensive dataset can be used to train the ML model, potentially leading to better performance and more accurate predictions.
4. Evaluate the performance of the ML model trained on the augmented dataset and compare it to the model trained on the original dataset.
5. Repeat 1 to 4 steps until we are satisfied with the model's performance.

Previous studies have shown that class-specific DA can improve the generalization performance of CNNs. The principal advantage of utilizing class-specific DA compared to global augmentation strategies is that the former facilitates a more individualized DA approach customized to each class in the dataset. Consequently, this may confer an edge regarding generalization capability and increased accuracy compared to a global DA approach [26]. This study focuses on automating the process of class-specific image DA in the context of multi-class classification. It utilizes a CNN as a tool for image analysis and a GA for searching for the best DA strategies.

Let us estimate the complexity of this problem. Let A be the set of all image transformations used in our study $A = \{a_1, \ldots, a_\alpha\}$. Here, α is the total number of such transformations. If the dataset contains l classes, the total number of possible class-specific DA transformations is α^l. Here we utilize Albumentations [4], a pre-existing library that provides access to more than $\alpha = 60$ diverse image transformation techniques. The datasets used in our study contain $l = 3$ and $l = 2$ distinct classes. Thus, the total size of the search space is $\alpha^l = 60^3 = 216,000$ and $60^2 = 3,600$, respectively.

Although this may not appear like a very large search space, the average time required to obtain a single solution is approximately 4.55 min, which involved training and testing a CNN on our system (Intel(R) Xeon(R) Platinum 8268 CPU 2.90 GHz, 132 GB of RAM, and a Quadro RTX 8000 GPU with 45556MiB using CUDA 11.2). It is easy to check that the total time required to test all possible solutions is approximately $216,000 \times 4.55\,\mathrm{m} \approx 1.87$years. Due to the enormous combinatorial challenge presented by the $216,000$ possible class-specific DA strategies and the associated time required to solve the problem (≈ 2 years), a naive approach is deemed impractical. In computational optimization, evolutionary algorithms have been widely acknowledged as efficient for large combinatorial problems, thus offering a feasible solution to this challenge [8].

To address this, we present a framework that integrates two crucial elements: a GA and a CNN, that function in a mutually beneficial fitness-score relationship to determine the optimal class-specific DA strategy. The GA plays a pivotal role in automating the selection of class-specific DA strategies required to augment the original dataset. The resulting augmented dataset is then utilized for training the CNN architecture, with the validation loss serving as the fitness score, guiding the GA's evolution process.

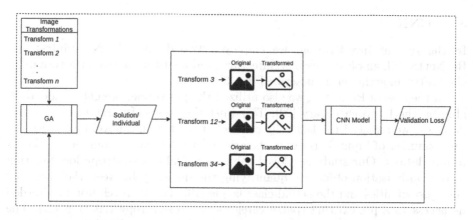

Fig. 1. The proposed framework starts from the GA generating a random population from a given list of image transformations. An individual contains l augmentations where l is the number of classes in the dataset (e.g. 3). The augmentations are used to double the size of the dataset. Then the dataset is fed into the CNN model, outputting a validation loss which we use to calculate and assign a fitness score to the individual.

The workflow of the proposed framework, as seen in Fig. 1, is composed of the following steps:

1. The CNN is trained on the original datasets. This process is repeated ten times, and the resultant baseline score is obtained by averaging the scores from these training sessions.
2. Next, the GA is employed to automatically generate a set of class-specific DA strategies, which we call solutions.
3. For each class-specific DA strategy (solution) produced by the GA, a new dataset is constructed by combining the baseline dataset with the augmented dataset, thus doubling the number of images.
4. The CNN is then trained on the augmented dataset, and metrics such as accuracy and loss are recorded.
5. The subsequent evaluation of the results involves comparing them with the average baseline score to determine the effectiveness of the proposed framework.

This framework offers a powerful solution for automating the selection of DA strategies and improving the performance of image classification models. The results of this study will contribute to a deeper understanding of the capabilities of this framework and inform future research in this area.

In the subsequent portion of this section, we will expound upon the CNN architectures used, datasets, pre-processing steps, and the representation of the GA and its operators.

3.1   CNN

In the present investigation, we employed two distinct CNN architectures: ResNet18 [14], an off-the-shelf architecture, and a customized architecture modelled after an architecture described in [23].

In the case of ResNet18, we also utilized the pre-trained weights made available by the PyTorch library [22]. To adjust the ResNet18 architecture to our datasets, we replaced the last fully connected layer with a novel layer with the same number of input features as the original model and the number of classes in our dataset. Our study employed the categorical cross-entropy loss function as our optimization objective, quantifying the disparity between the predicted class probabilities and the actual class labels [19]. The PyTorch library afforded us access to a pre-existing sparse categorical cross-entropy criterion [22]. The CNN architectures were optimized using the Adam and AdaDelta optimizers [17], which are well-suited for fine-tuning deep learning models and use gradient descent to update the model parameters, with additional features such as adaptive learning rates and momentum to improve convergence.

3.2   Datasets and Pre-processing

To guarantee the versatility of the proposed framework across diverse applications, we conducted experiments on three datasets originating from various domains, such as medicine and fashion e-commerce. The initial dataset employed in this study is MESO [7], which comprises tissue microarrays stained with hematoxylin and eosin, two commonly utilized staining agents to enhance the visualization of cells in biopsies. The dataset is composed of 243 cores, of which 155 are epithelioid, 64 are biphasic, and 24 are sarcomatoid in type. The second dataset employed in this study is the Fashion dataset, a subset of the DeepFashion2 dataset [11]. This dataset aims to distinguish between three comparable classes, specifically pants, joggers, and jeans. Additionally, to replicate a scarcity of data scenario, we randomly chose 300 images for each class, resulting in a total of 900 images for the complete dataset. The Breast Histopathology dataset, containing 277,524 images measuring 50×50 pixels, is the third dataset employed in this study, consisting of 198,738 IDC-negative images and 78,786 IDC-positive images. Invasive Ductal Carcinoma (IDC) is the most frequently occurring breast cancer subtype (Fig. 2).

Table 1. Number of images, classes and splits used in our experiments

Dataset	# Classes	# of samples	Split
MESO	3	243	80/10/10
Fashion	3	900	80/10/10
BreastHisto	2	58,742	70/15/15

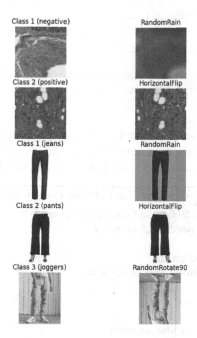

Fig. 2. Examples of original images (left) and augmented images (right).

The datasets were divided into the train, validation, and test subsets using random split, as seen in Table 1. The training subset was employed to train the model, updating the model parameters to minimize the loss function through the backpropagation of the optimisation algorithm (Adam or AdaDelta). Conversely, the validation subset was not utilized to update the model parameters but to evaluate the model's performance and prevent overfitting. The test subset was kept separate from the training and validation sets and only used once to evaluate the final model performance after the training process. To ensure compatibility with the ResNet18, all images were resized to a resolution of 224×224 pixels. The entire dataset underwent normalization with means $(0.485, 0.456, 0.406)$ and standard deviations $(0.229, 0.224, 0.225)$.

3.3   Genetic Algorithm

In this investigation, we employ a simple two-step genetic algorithm (GA) that utilizes a singular crossover operator and incorporates the application of elitism. The GA commences by producing a series of random solutions, subsequently utilized to expand the primary datasets and train the convolutional neural network (CNN). The minimum validation loss attained through CNN training is utilized to compute a fitness score subsequently assigned to each solution; see Fig. 3. The second step selects the best solutions for the crossover operation.

Fig. 3. Example of a GA-created solution and later scored by our CNN as found in our code, here F represents the fitness score.

During the crossover, each top solution is decomposed into its constituent transformations, which are then aggregated into a single list. New solutions are then formed randomly using this aggregated list; see Algorithm 2.

Algorithm 1 Genetic Algorithm for optimal class-specific DA

Require: A list of possible augmentations, target score T
Ensure: Optimal class-specific DA strategy
1: Initialize population
2: **while** not last generation **do**
3:    Evaluate each solution and assign a fitness score F ;
4:    **if** $F \geq T$ **then**
5:       *break*;
6:    **end if**
7:    Rank solutions from best to worst;
8:    Select top n solutions as parents;
9:    Keep the best solution intact (elitism);
10:    Create new offspring through crossover;
11:    Replace the old population with newly created offspring and the elite individual;
12: **end while**
13: **return** Best individual

The GA's evolution process, as seen in Algorithm 1, is guided by a fitness function that evaluates the performance of each solution and returns a fitness score F. This is calculated as the value of the inverse of the best validation loss achieved during the training of the CNN architecture, $F = |\text{ValidationLoss}|^{-1}$. This approach allows the GA to prioritize the solutions that lead to the lowest validation loss and, consequently, to the highest fitness score.

Algorithm 2 GA's crossover operator

Require: List of top n individuals;
Ensure: New offspring formed through crossover;
1: Combine all image augmentations from top individuals into a single list
2: Shuffle the list to introduce randomness;
3: Select a random number m of augmentations and form new offspring ;
4: **return** New offspring list;

---

4   Experimental Setup

In this research article, we conduct two experimental setups. The first setup involves replicating the experimental conditions described in [23], where we utilize the CNN and workflow presented by the authors along with k-fold cross-validation, elitism, dataset, and using Imgaug library and image transformations. K-fold cross-validation is a widely used technique in ML and statistical modelling to evaluate a predictive model's performance and generalization ability [1]. Specifically, k-fold cross-validation entails randomly partitioning the dataset into k equal-sized subsets or folds, followed by the iterative training and testing of the model k times. As a second experimental setup, we utilize only the genetic algorithm and the crossover operator without incorporating the elitism strategy and k-fold cross-validation. Here we use the Albumentations library.

In both setups, we train the CNN architectures on the datasets from Table 1 and record their baseline performances, saving the model with an R prefix followed by a unique code (e.g., R_j2763), so we can later differentiate it from the others, which are only saved as a unique code (e.g., 2nj942). We record metrics, such as training loss and accuracy, validation loss and accuracy, confusion matrices, precision, recall, the F_1-score, best epoch, total training time, fitness score and the augmentation strategy used. However, for the evaluation of the model, we only use the validation loss and fitness score as described in Sect. 4.1.

4.1   Evaluation Metric

Validation loss is a commonly used metric in the training of neural networks. It measures how well the model can generalize to new, unseen data. The validation loss is calculated using a separate data set held out during training, known as the validation set. This set evaluates the model's performance on data not seen during training. The validation loss is calculated after each epoch using the PyTorch built-in categorical cross-entropy [22]. The validation loss is calculated by feeding the validation set through the trained model and calculating the average error between the model's predictions and the actual class labels. It can be used to track the model's performance throughout training and to identify overfitting, which occurs when the model starts to memorize the training data and performs poorly on new data. In summary, the validation loss is an important metric for evaluating the performance of a neural network, especially for avoiding overfitting and ensuring that the model generalizes well to new, unseen data (Table 2).

Table 2. Performance of GA evolution. The numbers show an average cross-entropy loss, their differences and per cent improvement relative to a baseline.

TestID	Baseline	Avg10BestSol	Diff	(%)	Best Base	Best Sol	Diff	(%)
T_6b7326	.3780	.3676	.0104	(2.7%)	.3750	.3669	.0081	(2.2%)
T_347958	.4027	.3685	.0342	(8.5%)	.3951	.3586	.0365	(9.3%)
T_f24029	.6873	.6844	.0029	(0.4%)	.6864	.6843	.0020	(0.3%)
T_38b1e9	.6715	.6573	.0142	(2.1%)	.6696	.6549	.0147	(2.2%)

5   Results

In this section, we present the results of our experiments. Across all three datasets examined, our research consistently reveals notable improvements over the baseline. Notably, on the BreastHisto dataset, which already featured a substantial volume of images, we observed a commendable 8.49% improvement. In stark contrast, the MESO dataset, characterized by its comparably smaller and skewed nature, exhibited a significant performance boost, registering an impressive 18.99% improvement. The comparison between the baseline validation losses and the performance of the GA is presented in Table 3. The results indicate that the GA successfully discovered many class-specific DA solutions outperforming the baseline. The table also displays the average validation losses of these improved solutions and highlights the best validation loss achieved for each experiment.

In Experiment 1, the results do not consistently exhibit significant improvements. Nevertheless, the improvements are consistent even in cases where the CNN model fails to converge. Four tests were performed, and all showed improvements ranging from 2% to 5%. The failure of the CNN model to converge is not necessarily correlated with the framework but rather with the CNN architecture used. This observation is evident in T_f24029 and T_347958, where one of the main reasons for the difference between the average baselines (.6873 and .4027) could be the CNN architecture or the optimizer. Conversely, the average baselines are similar when the same CNN architecture is used, such as in T_347958 and T_6b7326 (.4027 and .3780). The results of T_6b7326 suggest that class-specific DA outperforms global DA, as the baseline training was conducted on a randomly oversampled dataset. The highest improvement ratio is observed in T_347958, with an improvement of 0.0342 over the baseline score.

The graphs presented in Fig. 4 and Fig. 5 illustrate the average fitness per generation, demonstrating the steady improvement of our Genetic Algorithm (GA) in enhancing the overall population fitness score over time. In test T_347958 Fig. 5, we observe the progress of a 5-generation GA, while in test T_6b7326, a 20-generation GA is displayed. In both cases, the populations evolve towards enhanced performance, but it's evident that increasing the number of generations results in more substantial performance improvements (Fig. 6).

In contrast, the results of Experiment 2 demonstrate more significant improvements, ranging from 0.1537 to 0.2616, representing a substantial improvement. We conducted four distinct experiments, all of which produced satisfactory outcomes. However, in T_e61c77, the dataset is exceptionally small, necessitating oversampling three or four times. Nevertheless, our framework could still identify better class-specific DA strategies than global DA. If the oversample baseline hyperparameter is true, it implies that the baseline validation was conducted on a randomly augmented dataset at the global level. We do this to simulate the manual selection of global DA and train the CNN on the same number of images.

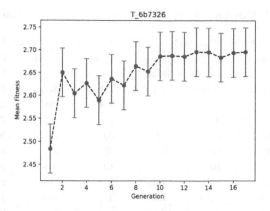

Fig. 4. Mean and STD of fitness over generations in Experiment 1 test T_6b7326.

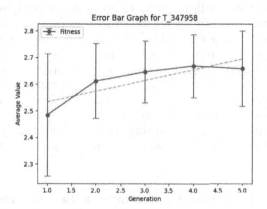

Fig. 5. Mean and STD of fitness over generations in Experiment 1 test T 347958.

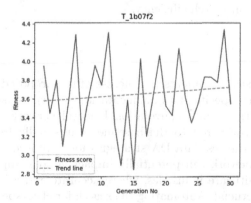

Fig. 6. Mean of fitness over generations in Experiment 2 test T_1b07f2.

Table 3. Experiment hyperparameters setup results, comparison

Hyper-parameter	T_6b7326	T_347958	T_f24029	T_7749c9	T_e61c77	T_fc8b83	T_1b07f2
Dataset	Breast	Breast	Breast	Fashion	MESO	Fashion	Fashion
CNN	ResNet18	ResNet18	EvoCNN	ResNet18	ResNet18	ResNet18	ResNet18
Optimiser	Adam	Adam	AdaDelta	Adam	Adam	Adam	Adam
Resolution	24	24	32	224	224	224	224
Population Coef	2	2	1	2	2	1	1
Generations	20	10	20	30	30	20	30
Top solutions	10	10	5	10	10	10	10
Offsprings	15	10	10	10	10	10	10
Oversample	True	False	False	True	True	False	False
Pre-trained	False	False	False	True	True	True	True
K-Fold	True	True	True	False	False	False	False
Experiment No	1	1	1	2	2	2	2
Data split	70/15/15	70/15/15	70/15/15	80/10/10	80/10/10	80/10/10	80/10/10
BaselineAvg	.3780	.4027	.6873	na	na	na	na
Avg10BestSol	.3676	.3685	.6844	.1718	.5773	.1677	.1481
Diff	.0104	.0342	.0029	na	na	na	na
(%)	(2.75%)	(8.49%)	(0.42%)	na	na	na	na
Best Baseline	.3750	.3951	.6864	.3639	.7127	.2968	.2944
Best Solution	.3669	.3586	.6843	.1426	.4511	.1431	.1102
Diff	.0081	.0365	.0020	.2213	.2616	.1537	.1842
(%)	(2.16%)	(9.23%)	(0.29%)	(60.81%)	(36.70%)	(51.79%)	(62.57%)
Time(h)	90.45	56.04	35.91	31.63	26.89	23.42	23.75

Finally, another substantial achievement is the time spent to find this improvement which does not exceed 3.75 days. Comparing this with the total time that would be needed to use the brute force approach 682.5 days results in 182% improvement in search efficiency.

6   Discussion

The study's conclusion highlights the success of the proposed automatic framework in finding class-specific data augmentation strategies using a combination of a genetic algorithm and a convolutional neural network. The improvement in validation loss compared to the baseline results indicates that the framework effectively optimizes data DA strategies for improved CNN performance. The proposed framework can potentially impact various applications in the e-commerce and healthcare industries, where accurate and robust image classification models are crucial. Automating the search for class-specific data augmentation strategies can save time and effort compared to manual experimentation with different augmentation strategies.

However, some limitations to the proposed framework should be considered. For instance, the framework may not be suitable for all data types and may require modifications for more complex and diverse data types. Furthermore, when working with medical images, it is imperative for medical specialists to validate the strategies discovered by the GA. In light of these limitations, future work should focus on extending the framework to handle more complex and diverse data types. This can help to improve the framework's performance and applicability to a broader range of image classification problems.

Overall, the proposed framework represents a promising approach to automating the search for class-specific data augmentation strategies, and the results of this study demonstrate its effectiveness in improving CNN performance.

7 Conclusion

In conclusion, this study introduces an automated framework to identify class-specific data augmentation strategies through a genetic algorithm and a convolutional neural network. The results demonstrate that the proposed framework successfully found superior class-specific data augmentation strategies compared to the manually set global augmentation strategy. The two experiments showed that the genetic algorithm was able to achieve improvements of 62.57% and 36.70% in validation loss in contrast to the baseline results. Moreover, a substantial accomplishment within this framework is the minimal time required to achieve these improvements, not exceeding 3.75 days. When juxtaposed with the total time that would have been necessitated by the brute force approach, amounting to a staggering 682.5 days, the efficiency of our framework shines. This comparison reveals an extraordinary 182% improvement in search efficiency.

These findings highlight the effectiveness of the proposed framework in enhancing CNN performance by optimizing data augmentation strategies. The proposed framework presents a promising approach for automating the search for class-specific data augmentation strategies that can benefit various industries, such as e-commerce and healthcare. Further research could extend the framework's capabilities to handle more complex and diverse data types while exploring different operators, such as controlled mutation rates for the genetic algorithm component.

References

1. Arlot, S., Celisse, A.: A survey of cross-validation procedures for model selection (2010)
2. Bäck, T., Schwefel, H.P.: An overview of evolutionary algorithms for parameter optimization. Evol. Comput. 1(1), 1–23 (1993)
3. Bharati, S., Podder, P., Mondal, M.R.H.: Hybrid deep learning for detecting lung diseases from x-ray images. Inform. Med. Unlocked 20, 100391 (2020)
4. Buslaev, A., Iglovikov, V.I., Khvedchenya, E., Parinov, A., Druzhinin, M., Kalinin, A.A.: Albumentations: fast and flexible image augmentations. Information 11(2) (2020). https://doi.org/10.3390/info11020125, https://www.mdpi.com/2078-2489/11/2/125

5. Cubuk, E.D., Zoph, B., Mané, D., Vasudevan, V., Le, Q.V.: Autoaugment: learning augmentation policies from data. CoRR abs/1805.09501 (2018). http://arxiv.org/abs/1805.09501

6. Dana, H., Ballard, C.M.B.: Computer Vision. Prentice-Hall, Hoboken (1982)

7. Eastwood, M., et al.: Malignant mesothelioma subtyping of tissue images via sampling driven multiple instance prediction. In: Michalowski, M., Abidi, S.S.R., Abidi, S. (eds.) Artificial Intelligence in Medicine. AIME 2022. LNCS, vol. 13263, pp. 263–272. Springer, Cham (2022). https://doi.org/10.1007/978-3-031-09342-5_25

8. Eiben, A.E., Smith, J.E.: Introduction to Evolutionary Computing. Springer, Berlin, Heidelberg (2015). https://doi.org/10.1007/978-3-662-43631-8_2

9. Erhan, D., Szegedy, C., Toshev, A., Anguelov, D.: Scalable object detection using deep neural networks. In: Proceedings of the IEEE Conference on Computer Vision and Pattern Recognition, pp. 2147–2154 (2014)

10. Frid-Adar, M., Diamant, I., Klang, E., Amitai, M., Goldberger, J., Greenspan, H.: Gan-based synthetic medical image augmentation for increased CNN performance in liver lesion classification. Neurocomputing **321**, 321–331 (2018). https://doi.org/10.1016/j.neucom.2018.09.013, https://www.sciencedirect.com/science/article/pii/S0925231218310749

11. Ge, Y., Zhang, R., Wu, L., Wang, X., Tang, X., Luo, P.: A versatile benchmark for detection, pose estimation, segmentation and re-identification of clothing images. In: CVPR (2019)

12. Goodfellow, I.J., Bengio, Y., Courville, A.: Deep Learning. MIT Press, Cambridge, MA, USA (2016). http://www.deeplearningbook.org

13. Hauberg, S., Freifeld, O., Larsen, A.B.L., III, J.W.F., Hansen, L.K.: Dreaming more data: class-dependent distributions over diffeomorphisms for learned data augmentation. CoRR abs/1510.02795 (2015). http://arxiv.org/abs/1510.02795

14. He, K., Zhang, X., Ren, S., Sun, J.: Deep residual learning for image recognition. CoRR abs/1512.03385 (2015). http://arxiv.org/abs/1512.03385

15. Hu, F., Xia, G.S., Hu, J., Zhang, L.: Transferring deep convolutional neural networks for the scene classification of high-resolution remote sensing imagery. Remote Sens. **7**(11), 14680–14707 (2015)

16. Khan, S., Javed, M.H., Ahmed, E., Shah, S.A.A., Ali, S.U.: Facial recognition using convolutional neural networks and implementation on smart glasses. In: 2019 International Conference on Information Science and Communication Technology (ICISCT), pp. 1–6 (2019). https://doi.org/10.1109/CISCT.2019.8777442

17. Kingma, D.P., Ba, J.: Adam: a method for stochastic optimization (2014). https://doi.org/10.48550/ARXIV.1412.6980, https://arxiv.org/abs/1412.6980

18. Kirkpatrick, S., Gelatt, C.D., Vecchi, J.M.P.: Optimization by simulated annealing. Science **220**(4598), 671–680 (1983)

19. Koidl, K.: Loss functions in classification tasks. School of Computer Science and Statistic Trinity College, Dublin (2013)

20. Krizhevsky, A., Sutskever, I., Hinton, G.E.: Imagenet classification with deep convolutional neural networks. Commun. ACM **60**(6), 84–90 (2017)

21. Nivin, T.W., Scott, G.J., Hurt, J.A., Chastain, R.L., Davis, C.H.: Exploring the effects of class-specific augmentation and class coalescence on deep neural network performance using a novel road feature dataset. In: 2018 IEEE Applied Imagery Pattern Recognition Workshop (AIPR), pp. 1–7 (2018). https://doi.org/10.1109/AIPR.2018.8707406

22. Paszke, A., et al.: Pytorch: an imperative style, high-performance deep learning library. CoRR abs/1912.01703 (2019). http://arxiv.org/abs/1912.01703

23. Pereira, S., Correia, J., Machado, P.: Evolving data augmentation strategies. In: Jiménez Laredo, J.L., Hidalgo, J.I., Babaagba, K.O. (eds.) EvoApplications 2022. LNCS, vol. 13224, pp. 337–351. Springer, Cham (2022). https://doi.org/10.1007/978-3-031-02462-7_22
24. Rebuffi, S., Gowal, S., Calian, D.A., Stimberg, F., Wiles, O., Mann, T.A.: Data augmentation can improve robustness. CoRR abs/2111.05328 (2021). https://arxiv.org/abs/2111.05328
25. S, A.K., Pal, A., Mopuri, K.R., Krishna Gorthi, R.: Adv-cut paste: semantic adversarial class specific data augmentation technique for object detection. In: 2022 26th International Conference on Pattern Recognition (ICPR), pp. 3632–3638 (2022). https://doi.org/10.1109/ICPR56361.2022.9956409
26. Shorten, C., Khoshgoftaar, T.: A survey on image data augmentation for deep learning. J. Big Data 6(1), 1–48 (2019)
27. Sonka, M., Hlavac, V., Boyle, R.: Image processing, analysis, and machine vision. Cengage Learning (2014)
28. Terauchi, A., Mori, N.: Evolutionary approach for autoaugment using the thermodynamical genetic algorithm. In: Proceedings of the AAAI Conference on Artificial Intelligence, vol. 35, pp. 9851–9858 (2021)
29. Ying, X.: An overview of overfitting and its solutions. J. Phys. Conf. Ser. 1168(2), 022022 (2019). https://doi.org/10.1088/1742-6596/1168/2/022022, https://dx.doi.org/10.1088/1742-6596/1168/2/022022

Augmented Lagrangian Method for Linear Programming Using Smooth Approximation

Hossein Moosaei[1](✉) [iD], Saeed Ketabchi[2] [iD], Mujahid N. Syed[3] [iD], and Fatemeh Bazikar[4] [iD]

[1] Department of Informatics, Faculty of Science, Jan Evangelista Purkyně University, Ústí nad Labem, Czech Republic
hmoosaei@gmail.com, hossein.moosaei@ujep.cz
[2] Department of Applied Mathematics, Faculty of Mathematical Sciences, University of Guilan, Rasht, Iran
sketabchi@guilan.ac.ir
[3] Department of Industrial and Systems Engineering, Interdisciplinary Research Center for Intelligent Secure Systems, King Fahd University of Petroleum and Minerals, Dhahran 31261, KSA, Saudi Arabia
smujahid@kfupm.edu.sa
[4] Department of Computer Science, Faculty of Mathematical Sciences, Alzahra University, Tehran, Iran
F.Bazikar@alzahra.ac.ir

Abstract. The augmented Lagrangian method can be used for finding the least $2-$norm solution of a linear programming problem. This approach's primary advantage is that it leads to the minimization of an unconstrained problem with a piecewise quadratic, convex, and differentiable objective function. However, this function lacks an ordinary Hessian, which precludes the use of a fast Newton method. In this paper, we apply the smoothing techniques and solve an unconstrained smooth reformulation of this problem using a fast Newton method. Computational results and comparisons are illustrated through multiple numerical examples to show the effectiveness of the proposed algorithm.

Keywords: Augmented Lagrangian method · Generalized Newton method · Smooth approximation

1 Introduction

In this paper, we consider linear program (LP) in the standard form

$$f_* = \min_{x \in X} c^T x, \quad X = \{x \in \mathbb{R}^n : Ax = b, \ x \ge 0\}, \tag{P}$$

where $A \in \mathbb{R}^{m \times n}, c \in \mathbb{R}^n$, and $b \in \mathbb{R}^m$ are given. In [12], Mangasarian, for solving the primal LP problem, proposed the usage of the exterior penalty function of

H. Moosaei et al. (Eds.): DIS 2023, LNCS 14321, pp. 186–193, 2024.
https://doi.org/10.1007/978-3-031-50320-7_13

the dual. The penalty function is piecewise quadratic, convex, and differentiable. Typically, the ordinary Hessian for this function does not exist everywhere. However, since its gradient is Lipschitz, a generalized Hessian exists [6,11,14,16,17]. These properties of the function allow him to establish a finite global convergence of the generalized Newton method.

In this study, we suggest the use of the augmented Lagrangian method as an alternative to the penalty function. We plan to implement smoothing techniques and solve a supplementary unconstrained smooth reformulation problem using the fast Newton method rather than the generalized Newton method [16,17]. The proposed approach can be applied to LPs with many nonnegative variables and a moderate number of equality-type constraints.

This paper is organized as follows. The augmented Lagrangian method is discussed in Sect. 2. The smooth approximation and a Newton-Armijo algorithm for solving the smooth reformulation problem are given in Sect. 3. In Sect. 4, numerical experiments on randomly generated problems are provided to illustrate the efficiency and validity of our proposed method. Concluding remarks are given in Sect. 5.

In this paper, all vectors will be column vectors and we denote the n-dimensional real space by $\mathbb{R}^n$. The notation $A \in \mathbb{R}^{m \times n}$ will signify a real $m \times n$ matrix. We mean A^T, $\|.\|$ and $\|.\|_\infty$ the transpose of matrix A, Euclidean norm, and infinity norm, respectively. The gradient $\nabla f(x)$ is a column vector. The gradient at the iterate x_k is $g_k = g(x_k)$ and the plus function x_+ is defined as $(x_+)_i = \max\{x_i, 0\}$, $i = 1, ..., n$.

2   Augmented Lagrangian Method

In this section, we briefly discuss applications of the augmented Lagrangian method to the LP (P). Assume that X_* and U_* are the solution set of the primal problem (P) and the solution set of its dual problem, respectively and $x \in \mathbb{R}^n$ is an arbitrary vector. The next theorem tells us that we can get a solution to the dual problem of (P) from an unconstrained minimization problem.

Theorem 1. *Assume that the solution set X_* of problem (P) is nonempty. Then there exists $\alpha_* > 0$ such that for all $\alpha \geq \alpha_*$ the unique least $2-$norm projection x of a point $\bar{x}$ onto X_* is given by $x = (\bar{x} + \alpha(A^T u(\alpha) - c))_+$ where $u(\alpha)$ is a point attaining the minimum in the following problem:*

$$\min_{u \in \mathbb{R}^m} \Phi(u, \alpha, \bar{x}) := -b^T u + \frac{1}{2\alpha} \| (\bar{x} + \alpha(A^T u - c))_+ \|^2 . \tag{1}$$

In addition, for all $\alpha > 0$ and $x \in X_*$, the solution of the convex, quadratic problem (1), $u_* = u(\alpha)$ is an exact solution of the dual problem i.e. $u(\alpha) \in U_*$.

Proof. Consider the problem of finding the least 2−norm projection $\bar{x}$ of the point x on X_*.

$$\min_{x \in X_*} \frac{1}{2} \parallel x - \bar{x} \parallel^2, \tag{2}$$

$$X_* = \{x \in \mathbb{R}^n : Ax = b, c^T x = f_*, \; x \geq 0_n\}.$$

The Lagrange function for the problem (2) is

$$L(x, u, \alpha, \bar{x}) = \frac{1}{2} \parallel \bar{x} - x \parallel^2 - u^T(Ax - b) + \alpha \, (c^T x - f_*),$$

where $u \in \mathbb{R}^m$, $\alpha \in \mathbb{R}$ are Lagrange multipliers and $\bar{x}$ is considered as a fixed parameter. The dual problem of (2) is

$$\max_{u \in \mathbb{R}^m} \max_{\alpha \in \mathbb{R}} \min_{x \in \mathbb{R}_+^n} L(x, u, \alpha, \bar{x}) \tag{3}$$

The optimality conditions of the inner minimization of the problem (3) is

$$\nabla L_x(x, u, \alpha, \bar{x}) = x - \bar{x} - A^T u + \alpha c \geq 0, \tag{4}$$
$$x^T(x - \bar{x} - A^T u + \alpha c) = 0, \quad x \geq 0. \tag{5}$$

It follows from (4) and (5) that the solution of this minimization problem is given in the following form:

$$x = (\bar{x} + A^T u - \alpha c)_+, \tag{6}$$

we replace x by $(\bar{x} + A^T u - \alpha c)_+$ into $L(x, u, \alpha, \bar{x})$ and obtain the dual function

$$S(u, \alpha, \bar{x}) = b^T u - \frac{1}{2} \parallel (\bar{x} + A^T u - \alpha c)_+ \parallel^2 - \alpha f_* + \frac{1}{2} \parallel \bar{x} \parallel^2 .$$

Hence formulation (2) is reduce to its dual problem

$$\max_{u \in \mathbb{R}^m} \max_{\alpha \in \mathbb{R}} S(u, \alpha, \bar{x}). \tag{7}$$

This problem is an optimization problem without any constraints, and its objective function contains an unknown value f_*. By [6–8] there exists a positive number $\alpha_* > 0$ such that, for each $\alpha > \alpha_*$, the projection x of the arbitrary vector $\bar{x} \in \mathbb{R}^n$ onto X_* can be obtained as following:

$$x = (\bar{x} + \alpha(A^T u(\alpha) - c))_+, \tag{8}$$

where $u(\alpha)$ is the solution of the problem (1), such that $\alpha \in \mathbb{R}$, $\alpha > \alpha_*$ is fixed. Gradient of $\Phi(u, \alpha, \bar{x})$ is

$$\nabla \Phi_u(u, \alpha, \bar{x}) = A(\bar{x} + \alpha(A^T u - c))_+ - b.$$

Assume $\bar{x} \in X_*$ and $\alpha > \alpha_*$ from (6) and by solving problem (1) we have

$$\bar{x} = (\bar{x} + \alpha(A^T u(\alpha) - c))_+, \tag{9}$$

it follows from (9)

$$\bar{x}^T(A^T u(\alpha) - c) = 0, \quad A^T u(\alpha) - c \leq 0. \tag{10}$$

This yields $u(\alpha)$ is a solution of the dual problem of (P), and the proof is complete.   □

We note that, the function $\Phi(u, \alpha, \bar{x})$ is the augmented Lagrangian function for dual problem of (P) (please see [1]),

$$f_* = \max_{u \in U} b^T u, \quad U = \{u \in \mathbb{R}^m : A^T u \leq c\} \tag{D}$$

The function $\Phi(u, \alpha, \bar{x})$ is piecewise quadratic, convex, and just has the first derivative, but it is not twice differentiable. Suppose that s and t are arbitrary points in $\mathbb{R}^m$. Then for $\nabla\Phi_u(u, \alpha, \bar{x})$ we have

$$\|\nabla\Phi_u(s, \alpha, \bar{x}) - \nabla\Phi_u(t, \alpha, \bar{x})\| \leq \|A\| \|A^T\| \|s - t\|,$$

this means $\nabla\Phi$ is globally Lipschitz continuous with constant $K = \|A\| \|A^T\|$. Thus, for this function, generalized Hessian exist and is defined by the $m \times m$ symmetric positive semidefinite matrix [9,10,12,13]. Now, we introduce the following iterative process :

$$u^{k+1} = arg \min_{u \in \mathbb{R}^m} \{-b^T u + \frac{1}{2\alpha} \| (x^k + \alpha(A^T u - c))_+ \|^2\}, \tag{11}$$

$$x^{k+1} = (x^k + \alpha(A^T u^{k+1} - c))_+, \tag{12}$$

where u^0 and x^0 are arbitrary starting point.

Theorem 2. *Let the solution set X_* of the problem (P) be nonempty. Then, for all $\alpha > 0$ and an arbitrary initial x^0 the iterative process (11), (12) converges to $x_* \in X_*$ in finite number of step k . The primal normal solution $\widehat{x}_*$ is obtained after the first iteration from the above process, i.e. $k = 1$. Furthermore, $u_* = u^{k+1}$ is an exact solution of the dual problem (D).*

The proof of the finite global convergence is given in [1].
In the next section, we describe an algorithm for solving the unconstrained optimization problem (1).

3   Smooth Reformulation

In this section, we develop an algorithm for solving an unconstrained minimization problem (11). This algorithm is based on smoothing methods presented
in [3,4]. Chen and Mangasarian [4] introduced a family of smoothing functions,
which is built as follows. Let $\rho : \mathbb{R} \to [0, \infty)$ be a piecewise continuous density
function satisfying

$$\int_{-\infty}^{+\infty} \rho(s)ds = 1, \ and \ \int_{-\infty}^{+\infty} |s|\rho(s)ds < \infty. \tag{13}$$

Then a smoothing function of the plus function is defined by

$$\varphi(x, \alpha) = \int_{-\infty}^{+\infty} max(0, s - \frac{x}{\alpha})\rho(s)ds. \tag{14}$$

Specific cases of this approach is,

$$\rho(s) = \frac{e^{-s}}{(1 + e^{-s})^2}, \quad \varphi(x, \alpha) = x + \frac{1}{\alpha}log(1 + e^{-\alpha x}). \tag{15}$$

The function φ with a smoothing parameter α is used here to replace the plus
function of (1) to obtain a smooth reformulation of problem (1):

$$\min_{u \in \mathbb{R}^n} f(u) := -b^T u + \frac{1}{2\alpha} \| (\varphi(x + \alpha(A^T u - c), \alpha) \|^2 . \tag{16}$$

Therefore, we have the following iterative process instead of (11) and (12).

$$u^{k+1} = arg \min_{u \in \mathbb{R}^n} \{-b^T u + \frac{1}{2\alpha} \| (\varphi(x^k + \alpha(A^T u - c), \alpha) \|^2,$$

$$x^{k+1} = (x^k + \alpha(A^T u^{k+1} - c))_+.$$

It can be shown that as the smoothing parameter α approaches infinity any solution of smooth problem (16) approaches the solution of the equivalent problem
(11) [4].
The twice differentiability of the objective function of the problem (16) allows
us to use a quadratically convergent Newton algorithm with an Armijo step
size [2,5]. This makes the algorithm globally convergent. We will now present a
Newton-Armijo algorithm for solving the problem (16).

In Algorithm 1, the Hessian may become singular; thus we use a modified
Newton direction as follows :

$$-(\nabla^2 f(u_i) + \gamma I_m)^{-1} \nabla f(u_i),$$

where δ is a small positive number ($\gamma = 10^{-4}$), and I_m is the identity matrix of
m order.

Algorithm 1. Newton's method with the Armijo rule.

Input: Choose any $u_0 \in \mathbb{R}^m$ and $tol > 0$;
　i=0, $s > 0$ be a constant, $\delta \in (0,1)$ and $\mu \in (0,1)$.
　while $\|\nabla f(u_i)_\infty\| \geq tol$,
　Choose $\alpha_i = max\{s, s\delta, s\delta^2, ...\}$ such that
　$f(u_i) - f(u_i + \alpha_i d_i) \geq -\alpha_i \mu \nabla f(u_i)^T d_i$,
　where $d_i = -\nabla^2 f(u_i)^{-1} \nabla f(u_i)$,
　$u_{i+1} = u_i + \alpha_i d_i$,
　$i = i + 1$;
　end

4   Numerical Results

In this section, we present some numerical results on various randomly generated problems to the problem (P). The problems are generated using the following MATLAB code (see Fig. 1):

```
1  % lpgen: Generate random solvable LP: min c'x s.t. Ax = b ;x
        >=0;
2  %Input: m,n,d(density); Output: A,b,c; (x, u): primal-dual
        solution;
3  m=input('Enter m: ');
4  n=input('Enter n: ');
5  d=input('Enter d: ');
6  pl=inline('(abs(x)+x)/2');
7  A=sprand(m,n,d);A=100*(A-0.5*spones(A));
8  x=sparse(10*(rand(n,1)));
9  u=spdiags((sign(pl(rand(m,1)-rand(m,1)))),0,m,m)*(rand(m,1)-
        rand(m,1));
10 b=A*x;c=A'*u+spdiags((ones(n,1)-sign(pl(x))),0,n,n)*10*ones(n
        ,1);
11 format short e;[norm(A*x-b), norm(pl(A'*u-c)), c'*x-b'*u]
```

Fig. 1. Code generation for random solvable LP.

The test problem generator generates a random matrix A and a vector b. The elements of A are uniformly distributed between -50 and $+50$. Our variant augmented Lagrangian method and the generalized Newton method and Armijo line search, were implemented in MATLAB. All computations were performed on *Core 2 Duo 2.53 GHz* computer with RAM of *4 GB*. We present a comparison between the proposed method, Smooth Augmented Lagrangian Newton (SALN) method and the **CPLEX** . All computations were conducted on the Windows operating system. The results of the numerical experiment are provided in Table 1. In all numerical examples, $tol = 10^{-10}$ and the starting vector is $x^0 = 0$. The total execution time for each method is given in the last column of the table. The accuracy of optimality conditions of linear programming problems are in the columns 4, 5, 6.

Table 1. Comparison of **SALN** and **CPLEX**.

m, n d	$Methods$	$\|x^*\|$	$\|Ax^* - b\|_\infty$	$|c^T x^* - b^T u^*|$	$\|(A^T u^* - c)_+\|_\infty$	$Time$ (sec)
500,600	SALN	$1.35e + 002$	$1.84e - 007$	$8.73e - 007$	$3.40e - 011$	4.77
1	CPLEX	$1.70e + 002$	$9.86e - 010$	$-2.91e - 010$	0	2.51
800,1000	SALN	$1.69e + 002$	$4.14e - 007$	$-6.34e - 007$	$6.18e - 011$	17.52
1	CPLEX	$2.17e + 002$	$3.06e - 009$	$3.49e - 010$	0	12.58
1500,2000	SALN	$2.31e + 002$	$1.09e - 007$	$-2.29e - 007$	$2.83e - 011$	9.85
0.1	CPLEX	$3.47e + 002$	$5.67e - 009$	$2.25e - 010$	0	60.01
2000,2000	SALN	$2.57e + 002$	$1.43e - 007$	$-3.75e - 008$	$3.59e - 011$	129.02
0.1	CPLEX	$2.57e + 002$	$1.11e - 008$	$-2.32e - 010$	0	57.06
2000,2500	SALN	$2.66e + 002$	$7.02e - 009$	$-4.18e - 010$	$4.75e - 012$	25.44
0.1	CPLEX	$3.71e + 002$	$9.05e - 008$	$-1.81e - 011$	0	53.15
100,1000000	SALN	$7.54e + 001$	$5.13e - 008$	$-5.43e - 007$	$1.13e - 013$	8.00
0.01	CPLEX	$1.18e + 005$	$4.75e - 011$	$1.01e - 008$	0	2.91
1000,1000000	SALN	$2.61e + 002$	$5.56e - 009$	$-8.78e - 008$	$1.13e - 013$	13.01
0.001	CPLEX	$8.85e + 005$	$9.41e - 010$	$-4.65e - 010$	0	2.71
100,10000000	SALN	$7.58e + 001$	$5.68e - 011$	$-1.96e - 010$	$5.68e - 014$	44.62
0.000001	CPLEX	$7.26e + 002$	$5.18e - 013$	$1.45e - 011$	0	21.94
1000,10000	SALN	$1.45e + 002$	$2.36e - 011$	$-2.91e - 010$	$4.96e - 013$	0.20
0.0001	CPLEX	$3.53e + 002$	$3.78e - 013$	$7.27e - 012$	0	0.24
2000,5000	SALN	$1.74e + 001$	$2.22e - 012$	$5.22e - 012$	$2.43e - 017$	0.09
0.000001	CPLEX	$1.74e + 001$	$2.84e - 014$	0	0	0.18
5000, 1000000	SALN	$3.67e + 001$	$1.66e - 011$	$2.34e - 011$	$6.39e - 014$	6.11
0.00000001	CPLEX	$3.67e + 001$	$4.93e - 014$	$2.27e - 013$	0	1.62
10000,1000000	SALN	$5.67e + 001$	$1.27e - 011$	$7.15e - 011$	$2.84e - 014$	6.90
0.00000001	CPLEX	$5.67e + 001$	$7.18e - 014$	$-1.59e - 012$	0	2.34
10000, 50000	SALN	$1.26e + 002$	$1.52e - 011$	$5.72e - 011$	$1.13e - 012$	0.48
0.000001	CPLEX	$1.26e + 002$	$2.23e - 013$	$-1.09e - 011$	0	0.22

5   Conclusion

The smooth reformulation process, based on the augmented Lagrangian algorithm, allows us to use a quadratically convergent Newton algorithm, which accelerates obtaining the normal solution of a linear programming problem. In this paper, the proposed method is compared with the CPLEX solver for solving LP. All numerical experiments were conducted on synthetic problems that were solved using SALN and CPLEX. Table 1 compares our algorithm with the **CPLEX** *v.12.1* solver for linear programming problems. As indicated in Table 1, the proposed algorithm successfully solves all the problems. The algorithm provides high accuracy and a solution with the minimum norm [15] in a reasonable time frame (refer to the third and last column of Table 1). This demonstrates the effectiveness and efficiency of our approach. Although CPLEX has low solution time in most of the cases (Ex. line 1, 2, 4, 6, 8, 10–12), efficient implementation of the proposed algorithm along with preprocessing may further speed up our algorithm.

References

1. Antipin, A.S.: Nonlinear programming methods based on primal and dual augmented Lagrangian. Institute for System Studies, Moscow (1979)
2. Armijo, L.: Minimization of functions having Lipschitz continuous first partial derivatives. Pac. J. Math. $16(1)$, 1–3 (1996)
3. Chen, C., Mangasarian, O.L.: Smoothing methods for convex inequalities and linear complementarity problems. Math. Program. $71(1)$, 51–69 (1995)
4. Chen, C., Mangasarian, O.L.: A class of smoothing functions for nonlinear and mixed complementarity problems. Comput. Optim. Appl. $5(2)$, 97–138 (1996)
5. Chen, X., Ye, Y.: On homotopy-smoothing methods for variational inequalities. SIAM J. Contr. Optim., 589–616 (1999)
6. Evtushenko, Y.G., Golikov, A.I., Mollaverdy, N.: Augmented Lagrangian method for large-scale linear programming problems. Optim. Methods Softw. $20(4–5)$, 515–524 (2005)
7. Evtushenko, Y.G., Golikov, A.I.: The augmented Lagrangian function for the linear programming problems. dynamics of non-homogeneous systems. In: Popkov, Y.S. (ed.) Russian Academy of Sciences Institute for System Analysis (1998)
8. Golikov, A.I., Evtushenko, Y.G.: Search for normal solutions in linear programming problems. Comput. Math. Math. Phys. $40(12)$, 1694–1714 (2002)
9. Hiriart-Urruty, J.B., Strodiot, J.J., Nguyen, V.H.: Generalized Hessian matrix and second-order optimality conditions for problems with $C^{1,1}$ data. Appl. Math. Optim. $11(1)$, 43–56 (1984)
10. Kanzow, C., Qi, H., Qi, L.: On the minimum norm solution of linear programs. J. Optim. Theory Appl. 116, 333–345 (2003)
11. Ketabchi, S., Behboodi-Kahoo, M.: Augmented Lagrangian method within L-shaped method for stochastic linear programs. Appl. Math. Comput. 266, 12–20 (2015)
12. Mangasarian, O.L.: A Newton method for linear programming. J. Optim. Theory Appl. 121, 1–18 (2004)
13. Mangasarian, O.L.: A finite Newton method for classification. Optim. Methods Softw. 17, 913–930 (2002)
14. Moosaei, H., Bazikar, F., Ketabchi, S., Hladik, M.: Universum parametric-margin ν-support vector machine for classification using the difference of convex functions algorithm. Appl. Intell. $52(3)$, 2634–2654 (2022)
15. Moosaei, H., Ketabchi, S., Jafari, H.: Minimum norm solution of the absolute value equations via simulated annealing algorithm. Afr. Mat. 26, 1221–1228 (2015)
16. Moosaei, H., Ketabchi, S., Razzaghi, M., Tanveer, M.: Generalized twin support vector machines. Neural Process. Lett. $53(2)$, 1545–1564 (2021)
17. Pardalos, P.M., Ketabchi, S., Moosaei, H.: Minimum norm solution to the positive semidefinite linear complementarity problem. Optimization $63(3)$, 359–369 (2014)

The Coherent Multi-representation Problem for Protein Structure Determination

A. Mucherino[1(✉)] and J-H. Lin[2]

[1] IRISA, University of Rennes, Rennes, France
antonio.mucherino@irisa.fr
[2] RCAS and NBRP, Academia Sinica, Taipei, Taiwan
jhlin@gate.sinica.edu.tw

Abstract. The Coherent Multi-representation Problem (CMP) was recently introduced and presented as an extension of the well-known Distance Geometry Problem (DGP). In this short contribution, we establish a closer relationship between the CMP and the problem of protein structure determination based on NMR experiments. Moreover, we introduce the concept of "level of coherence" for the several representations involved in the CMP, and provide some details about some ongoing research directions.

1 Introduction

Determining the three-dimensional conformation of biological molecules is of extreme importance in structural biology. These conformations can in fact reveal the potential functions that such molecules are in change of performing in living beings [3]. Moreover, they can also disclose additional information on how a given molecule can interact with particular chemical components, hence aiding at the development of new drugs [15]. Several approaches have been proposed in the past years for the determination of protein conformations, and in this work we focus on the one based on new introduced Coherent Multi-representation Problem (CMP) [16] where the data are obtained through experiments of Nuclear Magnetic Resonance (NMR) [2,20].

Historically, the determination of protein structures with the solution NMR technique relies mainly on the distance restraints, which are mainly derived from the cross peaks in NOESY spectra [9]. It usually requires a large number of distance restraints to obtain highly accurate protein structures, and the useful distance restraints come from the NMR signal from residues far from each other in the amino acid sequences of proteins. However, it is a highly non-trivial task to obtain a large number of distance restraints, due to the overlaps of resonances and peaks, artifacts in spectra and noise, and the missing signals due to the too fast motions of protein domains and loops. Once the collection of the distance restraints of the protein of interest is sufficient, a popular approach of determining protein structures are Simulated Annealing (SA) heuristics with constrained

H. Moosaei et al. (Eds.): DIS 2023, LNCS 14321, pp. 194–202, 2024.
https://doi.org/10.1007/978-3-031-50320-7_14

molecular dynamics simulations in the torsional space [10,21]. A key ingredient in this approach is an efficient algorithm for calculations of first and second derivatives of potential energy in the torsional space [1], which enables torsional angle dynamics (TAD) and its many application. More recently, NMR spectra are also exploited to find some estimations on the feasible values of torsion angles that can be defined on protein backbones [24].

By focusing mainly on the distance information, other approaches have been proposed in the scientific literature in the past years in order to overcome the typical issues related to the use of heuristic methods. Historical examples are the approaches in [4,5,11], while more recent efforts can be found for example in the survey [13] and the collection [17].

In the CMP, the main idea is to have different geometric representations for the atoms of the protein (for example, one based on distances, and another rather based on angles), that finally need to be "coherent" when constructing the overall three-dimensional conformation for the protein [16]. The CMP is NP-hard and it generalizes another problem known in the scientific literature as the Distance Geometry Problem (DGP) [13].

In this short contribution, we will make a point on the current ongoing research activities in the context of the CMP for the specific application arising in structural biology. Section 2 will briefly introduce the CMP in the context of the considered biological problem, by pointing out how NMR experiments are actually able to provide the exploited geometrical information. Then, we will discuss in Sect. 3 our current efforts in adding new and more complex representations to the CMP, while Sect. 4 some ideas will be presented for an automatic differentiation of the mathematical function in charge of measuring the level of coherence of each representation. Finally, Sect. 5 will briefly sum up the content of the paper, and conclude.

2   The CMP for Protein Structure Determination

Proteins are chains of amino acids. Each of the 20 amino acids that can be involved in the protein synthesis share a common group of atoms which takes part in forming the so-called protein backbone. The atoms contained in the protein backbone are basically hydrogen, carbon and nitrogen atoms. For simplicity in this work, we focus on protein backbones only.

A simple representation for a protein backbone consists in assigning a triplet of Cartesian coordinates to each of its atoms:

$$x : V \longrightarrow \mathbb{R}^3, \tag{1}$$

where V is a given set of atoms forming the protein. However, in structural biology, it is more common to employ another alternative representation for the protein backbones, which is based on the torsion angles that we can define for each quadruplet of consecutive atoms:

$$\omega : V \setminus \{v_1, v_2, v_3\} \longrightarrow [-2\pi, 2\pi), \tag{2}$$

where the numerical labels assigned to the vertices indicate their "natural order" in the protein, the one that follows the protein sequence and allows us to define a meaningful torsion angle for every quadruplet of consecutive vertices (these angles are generally referred to as *dihedral angles*). The first 3 vertices are not in the domain of ω, and we suppose that their Cartesian coordinates can be fixed, together with the other distances and angles involved in the definition of the torsion angles [14].

The study of the chemical composition of protein backbones can already give us some important information about protein conformations. Indeed, the presence of an atom in the protein backbone is known in advance, together with the information about its neighboring atoms through the existence of chemical bonds. In presence of a chemical bond, we can associate a distance to the two bonded atoms, which basically only depends on the kinds of atoms. Similarly, we can estimate in a rather precise way the distances between atoms that are chemically bonded to another common atom.

This overall distance information is very useful for protein modeling but it only gives local information: it gives no clues about the interactions between atoms that are located in parts of the chain that are separated by several amino acids in the protein sequence. Information with this global character can be obtained by performing experiments of Nuclear Magnetic Resonance (NMR) [2]. Generally, these experiments focus on hydrogen atoms and they can detect, by studying the magnetic properties of atomic nuclei, the close proximity of hydrogen atoms belonging to amino acids largely distant in the protein sequence. On the basis of this detection mechanism in NMR, it is possible to assign a distance value to pairs of hydrogen atoms that can be found in different parts of the protein, providing in this way information about the global geometry of the protein.

Several computational approaches are based on Euclidean distances only [17], i.e. geometrical information having a pure distance nature. However, distances are not the only kind of information that we can use in the context of protein structure determination. In order to introduce another kind of information, we point out that the NMR experiments are able to perform the prediction on the distances by studying some electro-magnetic signals that are related to the *precession* phenomenon of the aforementioned atomic nuclei in the proteins [20]. It turns out that this very same signal can also be used in artificial neural networks, such as the ones developed in tools of the TALOS *family* (TALOS+, TALOS-N) that exploit such raw signals to perform estimations on the typical torsion angles ϕ and ψ that we can define on protein backbones [23,24].

While some attempts of conversions from non-distance to distance information can be found in the scientific literature [12], it is evident that these conversions cannot always be performed without losing part of the information. For this reason, the CMP has been introduced in [16] in order to take into consideration different kinds of geometrical information at once in problems such as the one we are focusing in this paper in the context of structural biology.

The CMP can be represented by a simple directed graph $G = (V, E)$, where the vertices are atoms and there is an edge between two atoms when a geometrical relationship between them can be established. Orientations of the edges directly imply an order on the vertices of G. No weights are associated to the edges, nor to the vertices. It is rather through the specific multi-representation system that numerical values can be associated to the edges of the graph. In order to exploit the information obtained through NMR experiments, we can consider that one kind of information is given by the distance between two atoms, while another kind of information rather connects two atoms because there exists a torsion angle in the protein backbone having these two atoms at its two extreme ends.

The possibility to have multiple representations occurs at two different levels in the CMP. First of all, for every edge of the set E, we can consider more than one representation for the destination vertex (recall that the edges are directed in G). For example, for the same pair of vertices, we may have both distance and torsion angle information, hence allowing for the destination vertex to have two different representations. Moreover, the multiplicity of the representations can also be found again at the level of the entire graph, where all these local representations will have to find coherent values for their internal variables for defining a final and global realization x for the graph. For the formal definitions, the reader is referred to [16].

Let Y be the definition domain for the internal variables related to a non-Cartesian representation for a given vertex (e.g. the torsion angle representation, where Y has only one dimension). We refer to a coordinate transformation, capable to convert the coordinates from one system to another, as a triplet (P, Y, f), where P is the set of parameters (e.g. the fixed distances and angles in the quadruplets related to the torsion angles), Y is the coordinate domain, and f is the mapping:

$$f : (p, y) \in P \times Y \longrightarrow x \in \mathbb{R}^3, \tag{3}$$

that performs the transformation from the variables $y \in Y$ to the standard Cartesian coordinates $x \in \mathbb{R}^3$. The definition is given above for dimension 3 but can be trivially extended to any dimensions; we normally suppose that f is differentiable.

We say that a given vertex has *expected* Cartesian coordinates when its (possibly several) representations are all defined, so that the average value over each of its Cartesian coordinates can be computed. When these average values all correspond (for a given tolerance) to the ones obtained through the transformations, we say then that the several representations for the vertex are coherent. The *level of coherence* can be simply computed as the averaged variance among the different Cartesian coordinates obtained through the various representations for the same vertex. The level of coherence can play the role of penalty function in optimization problems attempting to find values for all representation variables that are able to make the system coherent.

3   New Representations in the Multi-representation

The *secondary structure* of a protein is related to some typical local arrangements of the protein chain. There are only two main secondary structures. The α-helix is an arrangement of the sequence of consecutive amino acids having the form of a helix where the number of amino acids per turn is fixed by chemical bonds so that the entire structure can only have slight fluctuations. The β-sheet structure is a little more complex: it is composed by a set of sequences of consecutive amino acids which take the form of a strand (where basically the amino acid appear like being aligned on an almost straight line). Different parts of the protein sequence can have the form of a strand, and then such a set of strands can have different possible ways to arrange themselves in the three-dimensional space in order to form a β-sheet. Because of the chemical bonds that take place between pairs of strands when generating a β-sheet structure, we can count combinatorially the various possible organizations in space for these strands.

When information about the protein secondary structures is available, we can consider to integrate it in our CMP instances. In previous works, this kind of information was translated into distance information [18], and subsequently included in the list containing the other available distances. In the context of the CMP, we can rather consider to have a dedicated kind of representation in the multi-representation system where several atoms are involved, and not only one per time.

From a mathematical point of view, the introduction of representations acting on multiple atoms requires an extension of the codomain of the transformation mappings involved in the change of coordinates (from internal coordinates for a given representation, to the standard Cartesian coordinates):

$$\hat{f} : (p, y) \in P \times Y \longrightarrow \hat{x} \in \mathbb{R}^{3 \times m},$$

where m is the number of involved atoms. Here atoms are not represented as single entities (as it is the case when using distance or angle information) but they are rather together with other atoms in some local conformations. From a technical point of view, this extension has an important impact on the current implementation in Java of the CMP[1]. In fact, the notification system, which attempts keeping all representation in a coherent state at all times, is currently based on the idea that, as soon as there is a change in one representation, all the others that depend on this one are notified of the change. When one of such notifications will arrive to a representation involving several atoms, the system will have to manage all these atoms, and possibly notify in turn all the other representations regarding these atoms.

An immediate consequence of including new representations based on secondary structures is therefore an expected higher complexity of the implementations. We expect however to strongly reduce the search space related to our CMP instances when these new representations will be taken into consideration.

---

[1] https://github.com/mucherino/DistanceGeometry, javaCMP folder.

4   Dual Numbers and Automatic Differentiation

Dual numbers remind of complex numbers for the way they are defined, but are essentially different [22]. They are pairs of real numbers (r, t), which we can also write as $r + \epsilon t$, where it is supposed that $\epsilon^2 = 0$. This assumption implies the definition of specific arithmetic operations on dual numbers where the component t is able to carry information about derivatives of functions $f(r)$ where r appears as a variable. When replacing a dual number $(r, 1)$ with the real number r in a given mathematical function, in fact, one simple function evaluation in $(r, 1)$ allows us to construct the new dual number $(f(r), f'(r))$, which contains the value of the function in r, as well as the value of the function derivative in r. We remark that the approach is also valid for trigonometric functions, and can be extended to functions admitting multiple variables.

In our ongoing research, the main idea is to use the properties of dual numbers to perform an automatic differentiation of the transformation functions (see Eq. (3)), in order to automatically construct the gradient of the function measuring the level of coherence for the multi-representations. We will give an example below related to the two representations via Cartesian coordinates in Eq. (1) and through torsion angles (see Eq. (2)). The importance of tools for automatic differentiation comes from the need of using local optimization methods, such as the Spectral Projected Gradient (SPG) [19] and the Levenberg-Marquardt Algorithm (LMA) [8], both already used in the context of distance geometry in previous publications.

A torsion angle ω_i corresponds to a quadruplet of Cartesian coordinates for consecutive vertices v_{i-3}, v_{i-2}, v_{i-1} and v_i. Since the very first 3 vertices are supposed to be fixed in the Cartesian space, every torsion angle ω_i only controls the position of the fourth vertex in each quadruplet of consecutive vertices. The transformation formula that we employ is the one proposed in [7], and it makes use of the three Cartesian coordinates for v_{i-3}, v_{i-2} and v_{i-1}, as well as of the value of ω_i, to compute the Cartesian coordinates for v_i. This is done by defining a local coordinate system having as origin the Cartesian coordinates of v_{i-1}. The unit vectors of the defined coordinate system can be organized in matrix form so that to define a 3×3 matrix to which we refer with the symbol U_{i-1} (our notations are consistent with the original discussion in [7]).

From a given torsion angle ω_i, the Cartesian coordinates (x_i, y_i, z_i) of v_i, together with the numerical derivatives (x_i', y_i', z_i'), over the three coordinates, of the applied transformation, can be computed as:

$$\begin{bmatrix} (x_i, x_i') \\ (y_i, y_i') \\ (z_i, z_i') \end{bmatrix} = \begin{bmatrix} x_{i-1} \\ y_{i-1} \\ z_{i-1} \end{bmatrix} + U_{i-1} \begin{bmatrix} -d(i-1, i) \cos \theta_i \\ d(i-1, i) \sin \theta_i \cos (\omega_i, 1) \\ d(i-1, i) \sin \theta_i \sin (\omega_i, 1) \end{bmatrix}, \qquad (4)$$

where θ_i is a constant and corresponds to the vector angle formed by the Cartesian coordinates of v_{i-3}, v_{i-2} and v_{i-1}. Notice the use of the dual number $(\omega_i, 1)$ as an argument for both sine and cosine functions. Naturally, the calculations need to be performed in dual arithmetic in order to obtain the desired result.

We are aware of the fact that the transformation mapping used above is rather simple, so that the symbolic differentiation may be used instead. If we opt for the use of automatic differentiation through dual arithmetic is because of our wish is to keep our method as general as possible. This will allow us to more easily integrate, for example, the new alternative representations briefly discussed in Sect. 3. Moreover, it is worth mentioning that, in general, every representation is based on the expected Cartesian coordinates of another, which need to be taken into consideration when differentiating, in particular when the derivatives are to be computed on the variables used in the representation that plays the role of reference. We believe the use of dual numbers will make our implementations more efficient, as well as elegant.

5   Conclusions

We have presented the newly introduced CMP in an informal way (as opposite to the formal introduction appeared in [16]) by taking as an example the natural application of the CMP in the context of structural biology. Moreover, we have given some emphasis to the two main research directions that we are currently investigating for the CMP. Firstly, we are studying the possibility of adding new representations in the multi-representation system of the CMP, including some representations where groups of atom are involved at once. We have considered representations where the several atoms take part in the definition of protein secondary structures, but a further extension to "protein blocks" is also possible [6]. Secondly, we are working on automatic tools for the differentiation of the transformation mappings, and hence for the function in change to compute the level of coherence for the several representations for a given atom. These research lines will be further investigated in the near future, and the corresponding codes (currently under development) will appear on the GitHub repository named "`DistanceGeometry`".

Acknowledgments. This work is partially supported by the international PRCI project MULTIBIOSTRUCT, co-funded by the ANR French funding agency (ANR-19-CE45-0019) and the National Science and Technology Council (NSTC) of Taiwan (MoST 109-2923-M-001-004-MY3). Most of the discussions, giving rise to some of the ideas presented in this paper, took indeed place during one of the visits of AM to Academia Sinica.

References

1. Abe, H., Braun, W., Noguti, T., Gō, N.: Rapid calculation of first and second derivatives of conformational energy with respect to dihedral angles for proteins general recurrent equations. Comput. Chem. **8**(4), 239–247 (1984)
2. Almeida, F.C.L., Moraes, A.H., Gomes-Neto, F.: An overview on protein structure determination by NMR, historical and future perspectives of the use of distance geometry methods. In: Mucherino, A., Lavor, C., Liberti, L., Maculan, N. (eds.) Distance Geometry, pp. 377–412. Springer, New York (2013). https://doi.org/10.1007/978-1-4614-5128-0_18

3. Berezovsky, I.N., Guarnera, E., Zheng, Z.: Basic units of protein structure, folding, and function. Prog. Biophys. Mol. Biol. **128**, 85–99 (2017)
4. Braun, W., Bösch, C., Brown, L.R., Ō Nobuhiro, G., Wüthrich, K.: Combined use of proton-proton overhauser enhancements and a distance geometry algorithm for determination of polypeptide conformations. Appl. Micelle-Bound Glucagon, Biochimica et Biophysica Acta − Protein Struct. **667**(2), 377–396 (1981)
5. Crippen, G.M., Havel, T.F.: Distance Geometry and Molecular Conformation. John Wiley & Sons, Hoboken (1988)
6. de Brevern, A.G., Valadié, H., Hazout, S., Etchebest, C.: Extension of a local backbone description using a structural alphabet: a new approach to the sequence-structure relationship. Protein Sci. **11**(12), 2871–2886 (2002)
7. Gonçalves, D.S., Mucherino, A.: Discretization orders and efficient computation of cartesian coordinates for distance geometry. Optim. Lett. **8**(7), 2111–2125 (2014)
8. Gonçalves, D.S., Mucherino, A.: A distance geometry procedure using the Levenberg-Marquardt algorithm and with applications in biology but not only. In: Rojas, I., Valenzuela, O., Rojas, F., Herrera, L.J., Ortuno, F. (eds.) Bioinformatics and Biomedical Engineering. Lecture Notes in Computer Science(), vol. 13347, pp. 142–152. Springer, Cham (2022). https://doi.org/10.1007/978-3-031-07802-6_13
9. Güntert, P., Buchner, L.: Combined automated NOE assignment and structure calculation with CYANA. J. Biomol. NMR **62**, 453–471 (2015)
10. Güntert, P., Mumenthaler, C., Wthrich, K.: Torsion angle dynamics for NMR structure calculation with the new program DYANA. J. Mol. Biol. **273**(1), 283–298 (1997)
11. Havel, T.F., Kuntz, I.D., Crippen, G.M.: The theory and practice of distance geometry. Bull. Math. Biol. **45**, 665–720 (1983)
12. Hengeveld, S.B., Malliavin, T., Liberti, L., Mucherino, A.: Collecting data for generating distance geometry graphs for protein structure determination. In: Proceedings of ROADEF23, Rennes, France, p. 2, (2023)
13. Liberti, L., Lavor, C., Maculan, N., Mucherino, A.: Euclidean distance geometry and applications. SIAM Rev. **56**(1), 3–69 (2014)
14. Malliavin, T.E., Mucherino, A., Nilges, M.: Distance geometry in structural biology: new perspectives. In: Mucherino, A., Lavor, C., Liberti, L., Maculan, N. (eds.) Distance Geometry, pp. 329–350. Springer, New York (2013). https://doi.org/10.1007/978-1-4614-5128-0_16
15. Moreira, I.S., Fernandes, P.A., Ramos, M.J.: Protein-protein docking dealing with the unknown. J. Comput. Chem. **31**(2), 317–342 (2010)
16. Mucherino, A.: The coherent multi-representation problem with applications in structural biology. In: Rojas, I., Valenzuela, O., Rojas Ruiz, F., Herrera, L.J., Ortuno, F. (eds.) Bioinformatics and Biomedical Engineering. Lecture Notes in Computer Science(), vol. 13919, pp. 338–346. Springer, Cham (2023). https://doi.org/10.1007/978-3-031-34953-9_27
17. Mucherino, A., Lavor, C., Liberti, L., Maculan, N. (eds.): Distance Geometry: Theory Methods and Applications, p. 410. Springer, Cham (2013)
18. Mucherino, A., Lavor, C., Malliavin, T., Liberti, L., Nilges, M., Maculan, N.: Influence of pruning devices on the solution of molecular distance geometry problems. In: Pardalos, P.M., Rebennack, S. (eds.) Experimental Algorithms. Lecture Notes in Computer Science, vol. 6630, pp. 206–217. Springer, Berlin (2011). https://doi.org/10.1007/978-3-642-20662-7_18
19. Mucherino, A., Lin, J.-H., Gonçalves, D.S.: A coarse-grained representation for discretizable distance geometry with interval data. In: Rojas, I., Valenzuela, O.,

Rojas, F., Ortuño, F. (eds.) IWBBIO 2019. LNCS, vol. 11465, pp. 3–13. Springer, Cham (2019). https://doi.org/10.1007/978-3-030-17938-0_1

20. Nascimento, C.: Ressonância Magnética Nuclear (in Portuguese), p. 120. Blusher, São Paulo (2016)

21. Nilges, M.: Calculation of protein structures with ambiguous distance restraints. automated assignment of ambiguous NOE crosspeaks and disulphide connectivities. J. Mol. Biol. **245**(5), 645–660 (1995)

22. Pennestrì, E., Stefanelli, R.: Linear algebra and numerical algorithms using dual numbers. Multibody Sys. Dyn. **18**, 323–344 (2007)

23. Shen, Y., Bax, A.: Protein structural information derived from NMR chemical shift with the neural network program TALOS-N. Methods Mol. Biol. **1260**, 17–32 (2015)

24. Shen, Y., Delaglio, F., Cornilescu, G., Bax, A.: TALOS+: a hybrid method for predicting protein backbone torsion angles from NMR chemical shifts. J. Biomol. NMR **44**(4), 213–236 (2009)

The Effects of Shift Generation on Staff Rostering

Kimmo Nurmi[1](✉) [iD], Jari Kyngäs[1], and Nico Kyngäs[2]

[1] Satakunta University of Applied Sciences, Satakunnankatu 23, 28130 Pori, Finland
cimmo.nurmi@samk.fi
[2] University of Turku, 20014 Turun yliopisto, Finland

Abstract. To the best of our knowledge, this is the first paper to examine the effect of shift structures on staff rostering optimization. The study showed that the generated shift structures have a significant effect on optimizing the staff rosters. The study also showed that we should allow longer working days, because they imply better consideration of the stress and risk factors introduced by the Finnish Institute of Occupational Health. The PEASTP metaheuristic, a computational intelligence framework, was used to justify the findings. The results were obtained using a real-world instance from a Finnish contact center.

Keywords: Workforce Scheduling · Staff Rostering · Shift Generation · Metaheuristics

1 Introduction to Workforce Optimization

Workforce scheduling is a difficult and time-consuming problem that every company or institution that has employees working on shifts or on irregular working days must solve. Such business areas and organizations include hospitals, retail stores, call centers, cleaning, home care, guarding, manufacturing and delivery of goods. Labor is the largest expense for these organizations.

The main goal of workforce optimization is the performance of staff on financial efficiency. Other important goals include fairer workloads and employee satisfaction. Furthermore, the optimization process should address the health, safety, and well-being of the employees. The best optimization practices will help move an organization forward successfully. Workforce optimization is not a static action, but ongoing activity that helps an organization to become more efficient and effective as it grows. When an organization grows, having enough people to manage the work effectively becomes a major area of concern. Overtime is expensive, and so is to hiring additional full-time employees.

The ideal workforce scheduling process starts from *workload prediction* (see Fig. 1), which is done based on both known and predicted events. For example, the arrivals of customers can be predicted using patient flow in hospital's intensive care unit, cash receipts in supermarket, and received calls in contact centers. Known events may also be gathered from current sales contracts.

H. Moosaei et al. (Eds.): DIS 2023, LNCS 14321, pp. 203–217, 2024.
https://doi.org/10.1007/978-3-031-50320-7_15

The process continues with *shift generation*. The most important optimization target is to match the shifts to the predicted workload as accurately as possible. The generation of shifts is based on either the number of employees required to work at given timeslots or the tasks that the shifts have to cover. In some lines of businesses, such as in airlines, railways and bus companies, the demand for employees is quite straightforward because the timetables are known beforehand and the shifts are already fixed.

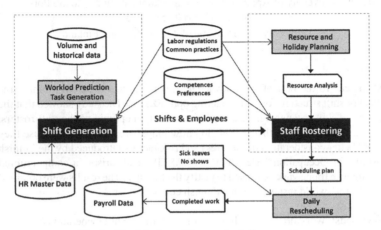

Fig. 1. The ideal workforce scheduling process.

The generated shifts form an input for *staff rostering*, where employees are assigned to the shifts. The most important optimization target is to guarantee a minimum number of employees with particular competences for each shift. The collective labor agreements and government regulations must be respected.

After the shift generation and staff rostering have completed, the optimized staff rosters need to be changed. This is due to sick leaves and other no shows, or because new tasks will arise and some of the tasks need to be changed or removed. *Daily rescheduling* deals with the ad hoc changes. The workforce management system should recommend suitable substitutes considering the qualifications, employment contract, legal limitations and salaries. The goal is to find the most economical candidates.

Future staffing requirements must be carefully considered in *resource and holiday planning* phase. Holidays, training sessions and other short-term absences as well as long-term sick-leaves and forthcoming retirements have major impact to actual staff rostering.

The ideal workforce scheduling process relies on both optimization resources and human resources. It links the organization together, optimizing processes and streamlining decision-making. When the input data for the workload prediction, shift generation and staff rostering phases are well validated, significant benefit in financial efficiency and employee satisfaction can be achieved.

The main contribution of this paper is to show that the generated shift structures have a significant effect on optimizing the staff rosters. The paper is organized as follows.

Section 1.1 introduces various aspects of shift generation. Section 2 gives the formulation of the real-world employee-based shift generation problem and the real-world staff rostering problem. In Sect. 3, we briefly describe the computational intelligence method used to solve the problems. Section 4 first describes the real-world instances used for the optimizations. Then we give the computational results to show the substantial consequences of the shift generation phase. Conclusion section summarizes our findings.

1.1 Introduction to Shift Generation

The generation of shifts is based on either the varying number of required employees working during the planning horizon or the tasks that the shifts must cover. We call these *employee-based shift generation* and *task-based shift generation* problems. The first major contribution for the employee-based shift generation problem was the study by Musliu et al. [1]. They introduced a problem, in which the workforce requirements for a certain period of time were given, along with constraints about the possible start times and the length of shifts, and an upper limit for the average number of duties per week per employee. Di Gaspero et al. [2] proposed a problem in which the most important issue was to minimize the number of different kinds of shifts used. The problem statement also includes a collection of acceptable shift types, each of them characterized by the earliest/latest start time and minimum/maximum length of its shifts.

Kyngäs et al. [3] introduced the unlimited shift generation problem in which the most important goal is to minimize understaffing and overstaffing. They define a strict version of the problem, in the sense that each timeslot should be exactly covered by the correct number of employees. The lengths and the start times of the shifts may vary. In the person-based multitask shift generation problem with breaks presented in [4], employees can have their personal shift length constraints and competences. The goal is to ensure that the employees can execute the shifts later in the staff rostering phase.

In the task-based shift generation problem the goal is to create shifts and assign tasks to these shifts so that the employees can be assigned to the shifts. The first major contribution of the task-based problem was the study by Dowling et al. [5]. They developed a day-to-day planning tool and to estimate a minimal staff set capable of operating as the ground staff of an international airport. Their two-stage approach employed a simulated annealing algorithm. In the first stage, worker shifts were derived over a two-week planning horizon, and in the second stage, tasks were allocated to shifts/workers. Valls et al. [6] introduced a model where they minimized the number of workers required to perform a machine load plan. They presented a coloring approach to identify possible allocations along with bounds on the branch-and-bound search tree.

Krishnamoorthy and Ernst [7] introduced a similar group of problems, which they called Personnel Task Scheduling Problems (PTSP). Given the staff that are rostered on a particular day, the PTSP is to allocate each individual task, with specified start and end times, to available staff who have skills to perform the task. Later, Krishnamoorthy et al. [8] introduced a special case referred as Shift Minimization Personnel Task Scheduling Problem (SMPTSP) in which the goal is to minimize the number of employees used to perform the shifts. The SMPTSP has been studied under a few other names. Jansen [9] called SMPTSP the license and shift class design problem. Kroon et al. [10] called

SMPTSP tactical fixed interval scheduling problem, and showed that solving it to optimality is NP-hard. The SMPTSP is also similar to the basic interval scheduling problem presented in [11] where the goal is to decide which jobs to process on which machines. Each machine has a given known interval during which it is available, and each job has an ideal machine on which it is preferred to be processed.

The General Task-based Shift Generation Problem (GTSGP) was defined in [12]. Given the tasks that should be rostered on a particular day, the GTSGP is to create anonymous shifts and assign tasks to these shifts so that employees can be assigned to the shifts. The targeted tasks must be completed within a given time window. For example, shelving in retail stores is often carried out in the forenoon. Some tasks are so-called back-office tasks. For example, in a contact center answering emails might require a given number of working hours per day dedicated to the activity but these tasks can be carried out any time of the day.

Modifying GTSGP to allow only fixed timetables for the tasks, we obtain the Extended Shift Minimization Personnel Task Scheduling Problem (ESMPTSP) defined in [13]. The problem is also an extension to the Shift Minimization Personnel Task Scheduling Problem. The objective is to first minimize the number of employees required to carry out the given set of tasks and then to maximize the number of feasible (shift, employee) pairs.

2   Employee-Based Shift Generation Problem and Staff Rostering Problem

In theory, the best practical results can be achieved when shift generation and staff rostering are processed and solved at the same time. However, different variations of both problems and even their sub-problems are known to be NP-hard and NP-complete [14–18]. Nonetheless, some interesting implementations exist. Jackson et al. [19] presented a very simple randomized greedy algorithm that uses very little computational resources. Lapegue et al. [20] introduced the Shift Design and Personnel Task Scheduling Problem with Equity objective (SDPTSP-E) and built employee timetables by fixing days-off, designing shifts and assigning fixed tasks within these shifts. They minimized the number of tasks left unassigned. Dowling et al. [20] first created a master roster, a collection of working shifts and off shifts, and then allocated the tasks in their Task Optimiser module.

Prot et al. [21] proposed a two-phase approach consisting in first computing a set of interesting shifts, then, each shift is used to build a schedule by assigning tasks to workers, and then iterating between these two phases to improve solutions. They relaxed the constraint that each task has to be assigned. Smet et al. [22] presented the Integrated Task Scheduling and Personnel Rostering Problem, in which the task demands and the shifts are fixed in time. Due to the complexity issues in large-scale practical applications, shift generation and staff rostering are often solved separately. Finally, Nurmi et al. [12] defined the General Task-based Shift Generation Problem (GTSGP), in which the task is to create anonymous shifts and assign tasks to these shifts so that employees can be assigned to the shifts. The goal is to maximize the number of shifts employees are able to execute. This should enhance the solutions generated by the staff rostering phase.

In the real-world workforce scheduling process, the employee-based shift generation phase is processed first, followed by the staff rostering phase. The shift generation problem for the real-world instance used in Sect. 4 is given as follows using the formulation in [3] and [4]:

C1.The sum of working time in the generated shifts must match the sum of the demand for labor as closely as possible.
C2.Excess in shifts (over demand) is minimized.
C3.Shortage in shifts (under demand) is not allowed.
V1.The number of different shifts is minimized
V3.Shifts of less than 9 hours (8 for Friday) and over 12 hours are minimized.
V4.The target for average shift length is 10 hours.
P1.Shifts cannot start between 22:01 and 06:59.
P2.Night shifts must end before 08:01.

Once we have generated the shifts, we continue with the staff rostering phase by assigning the employees to the shifts. From an employee perspective, shift work is associated with pressure on social and family life, health issues, motivation and loyalty. Humans live in a rhythm, in a biological rhythm and in a social rhythm. These rhythms and irregular shift work are undoubtedly in conflict. This can damage humans' health and well-being. Some individuals are more influenced by these effects than others are. Therefore, it is extremely important that employees have control over their working hours. This buffers the negative impact of shift work on their health and work-life balance. This understandably decreases sickness absences, in-creases working capacities and lengthens careers.

The Finnish Institute of Occupational Health (FIOH), which operates under the Ministry of Social Affairs and Health, published their recommendations for shift work [23] in May 2019. They presented twenty-two risk and stress factors grouped under five headings: length of the workday, timing of the working hours, recovery and rest times, reconciling work and family life and possibilities to influence personal working times. The institute introduced traffic lights for these factors to indicate the degree of overload.

The recommendations of FIOH are based on the ten-year longitudinal study between 2008 and 2018 (see e.g. [24] and [25]). The study followed the effects and con-sequences of shift work on 13 000 hospital workers. Based on the extensive cooperation, the institute has daily working time statistics of more than 150 000 employees working in healthcare and social services sector.

Table 1 shows the fourteen quantitative risk and stress factors presented by FIOH. The table shows the FIOH's recommendations for shift work when the scheduling timeframe is three weeks. Based on their scientific data on health and recovery issues, the three most important factors to consider are the number of night shifts, the rest time between two shifts and the longest work period between two days-off. The importance of these factors is denoted by one in Table 1.

The second most important factors (denoted by two) are long enough days-off, the number of free weekends, the number of night shifts in a row and the length of the shifts. Furthermore, it is important that an employee has an influence on her/his personal working times, i.e. she/he can make shift and days-off requests. The rest of the issues to consider are the number of working days in a row, the number of evening shifts in a row, the number of single working days, the longest rest period between two shifts and the rest time after a night shift.

Table 1. Fourteen stress and risk factors and the recommendations for shift work published by the Finnish Institute of Occupational Health for a planning horizon of three weeks.

Recommendations Overload → Importance ↓		Heavy (red)	Overload (orange)	Increased (yellow)	Safe (green)
1. Length of the workday					
A. Longest work period[1]	1	> 55h	> 48h	> 40h	≤ 40h
B. Working hours in longest shift[2]	2	> 14h	> 12h	> 10h	≤ 10h
C. Consecutive working days[3]	3	≥ 8	7	6 or 2	3-5 (all)
D. Single working days[3]	3	≥ 1			0
2. Timing of the working hours					
A. Consecutive evening shifts[3]	3	≥ 6	5	4	0-3 (all)
B. Consecutive night shifts	2	≥ 6	5	3-4	0 or 2 (all)
C. Single night shifts	2	> 0			0
D. Number of night shifts	1	≥ 9	5-8	3-4	0 or 2 (all)
3. Recovery and rest times					
A. Rest times below 11h[3]	1	≥ 9	5-8	2-4	0-1 (all)
B. Rest time after a night shift	3	< 11h	< 28h	≤ 48h	> 48h
C. Longest rest period[4]	3	< 24h	< 35h	< 48h	≥ 48h
4. Reconciling work and family life					
A. Free weekends[5]	2		0	1	2-3
B. Single days-off	2	≥ 4	3	2	0-1
5. Possibilities to influence personal working times					
A. Employees can make requests	2	No			Yes

[1]between two days-off, [2]shift length should not be under 4h, [3]at least one such case, [4]between two shifts, [5]both Saturday and Sunday (Sat 00:00 – Mon 00:00) free

Our recent work [26] showed that problems arise when all these recommendations should be satisfied together in real-world staff rostering. The interpretations showed the multifactorial and interdependent relations of the optimization targets, i.e. the targets compete with each other. A small change in one target may lead to inconsistency in another target. On the other hand, a small relaxation in one target may lead to substantial improvement in another target. Therefore, to consider all requirements, requests and interactions is well beyond the reach of human intelligence.

We used a computational intelligence framework to show how the recommendations and employer requirements compete with each other. Based on our studies, we published our practical recommendations for shift work shown in Table 2. The practical recommendations are such that a sophisticated metaheuristic can generate high-quality staff rosters so that the computation time is still acceptable considering the release time of the rosters.

Our study showed that the most challenging factors to optimize are not the same as the most important factors according to FIOH. For example, guaranteeing no single working days and no single days-off is extremely challenging. Furthermore, guaranteeing at least two free weekends is practically impossible. We also recommended not using single night shifts.

The optimization task of the staff rostering is two-fold. First, the solution should respect the basic requirements:

R1. A minimum number of employees with particular competences must be guaranteed for each shift.
R2. An employee can only be assigned to a shift she/he has competence for.
R3. Working Hours Act, Employment Contract Acts and practices at the organizational level must be respected.

Second, the solution should optimize the following:

O1. An employee's total working time should be within the given limits.
O2. The fifteen factors given in Table 1 should be considered.

Table 2. The practical recommendations for shift work.

RECOMMENDATIONS	IDEAL (FIOH)	OUR PRACTICAL ONES
1. Length of the workday		
A. Longest work period	≤ 40h	≤ 40h with low importance
B. Working hours in longest shift	≤ 10h	≤ 12h
C. Consecutive working days	3-5	2-6 (must be realized)
D. Single working days	Not allowed	Minimize with low importance
2. Timing of the working hours		
A. Consecutive evening shifts	0-3	0-4 with low importance
B. Consecutive night shifts	0 or 2	0 or 2-4 with high importance
C. Single night shifts	Not allowed	Not allowed (must be realized)
D. Number of night shifts	0-2	0 or 2-4 with high importance
3. Recovery and rest times		
A. Rest times below 11h	0-1	0-1 with low importance
B. Rest time after a night shift	> 48h	≥ 48h with high importance
C. Longest rest period	≥ 48h	≥ 35h (must be realized)
4. Reconciling work and family life		
A. Free weekends	2-3	1-3 (must be realized)
B. Single days-off	0-1	Minimize with low importance
5. Possibilities to influence personal working times		
A. Employees' prioritized requests	Yes	All must be realized
Employees' other request	Yes	Maximize with medium imp

In the context of the real-world instance described in Sect. 4, the optimization task is given as follows using the formulation in [27]. The timeframe is three weeks.

C1. An employee's shifts must not overlap.
C2. All the generated shifts must be allocated.
R1. An employee's mandatory limits for working hours and working/off days must be respected.
R3. At least one free weekend per employee is guaranteed.

R5.11 hours of rest time is guaranteed between adjacent shifts of an employee.

R5b.48 hours of rest time after a night shift is guaranteed.

R7.Employees work consecutively at most four days.

R7b.Employees work consecutively at most four evening shifts.

R10.A work period between adjacent days-off should be at most 40 hours.

R11.An employee should have at least one rest period of at least 35 hours each week.

O1.An employee must have all the competences required for their shifts.

O5.An employee cannot carry out particular shift combinations on adjacent days.

O8.Total working time for each employee should be within ±4 hours from their personal target.

E1.Single days-off should be minimized.

E2.Single working days should be minimized.

E8.An employee cannot carry out particular shifts immediately after absent days.

E10.An employee's maximum number of night shifts is four.

E11.Single night shifts are not allowed.

P2.Assign a shift to an employee on a day with a requested day-on. Assign no shift to an employee on a day with a requested day-off.

P3.Assign a requested shift to an employee on day with a shifts request. Do not assign a requested shift to an employee on a day with a request to avoid a shift.

3 The PEASTP Metaheuristic

A metaheuristic offers several tools that are problem independent and could be used as a skeleton for the development of a heuristic. The following six metaheuristics have indisputably introduced true novelties to the repertoire of optimization methods: simulated annealing [28], tabu search [29], genetic algorithm [30], variable neighborhood search [31], ruin and recreate method [32, 33] and ejection chain method [34].

By combining and carefully tuning the most efficient operators of these six metaheuristics, an experienced heuristic designer can solve real-life optimization problems efficiently. The PEASTP metaheuristic [35] is a good example of such combining and tuning. The metaheuristic has been in successful commercial use for several years, see e.g. [36] and [37].

The pseudo-code of the PEASTP metaheuristic is given in Fig. 2. The heuristic uses a *population* of solutions in each iteration. The worst solution is replaced with the best one at given intervals, i.e. elitism is used. The ejection chain *local search* explores promising areas in the search space. The ejection chain search is improved by introducing a *tabu list*, which prevents reverse order moves in the same sequence of moves. A *simulated annealing refinement* is used to decide whether to commit to a sequence of moves in the ejection chain search.

The simulated annealing refinement and tabu search are used to avoid staying stuck in promising search areas too long. *Shuffling* operators assist in escaping from local optima. They are used to perturb a solution into a potentially worse solution in order to escape from local optima. A shuffling followed by several ejection chain searches obtains better solutions using the same idea as the ruin and recreate method. The heuristic uses a penalty method, which assigns dynamic weights to the hard constraints based on the constant weights assigned to the soft constraints.

```
Set iteration limit t, population size n,
    elitism interval e, shuffling interval s and ADAGEN update interval a
Generate a random initial population of solutions Sᵢ for 1 <= i <= n
Set best_solution = null, iteration = 1
WHILE iteration ≤ t
    pop = 1
    WHILE pop <= n
        Apply ejection chain search to solution Sₚₒₚ to get a new solution
        IF Cost(Sₚₒₚ) < Cost(best_solution) THEN Set best_solution = Sₚₒₚ
        pop = pop + 1
    END WHILE
    Update simulated annealing framework
    IF round ≡ 0 (mod a) THEN Update the ADAGEN framework
    IF round ≡ 0 (mod s) THEN Apply shuffling operators
    IF round ≡ 0 (mod e) THEN Replace the worst solution with the best one
    Set iteration = iteration + 1
END WHILE
Output best_solution
```

Fig. 2. The pseudo-code of the PEASTP metaheuristic.

Each of the six metaheuristic components of PEASTP are crucial to produce good-quality solutions. This was verified for three different problem domains in [35]. Recently, the results in solving the General Task-based Shift Generation Problem [38] showed that when any one of the components was removed, the results were clearly worse. The same held even if the metaheuristic was given twice as much time to run without one of the components.

4 Computational Results

This section gives the computational evidence of the significant effects of shift structures on staff rosters. The study concerns an instance from a contact center. In the context of the instance, Sect. 2 described the detailed formulations and optimization targets of the shift generation and the staff rostering.

The PEASTP metaheuristic described in Sect. 3 is used to generate optimized shifts and optimized staff rosters. The metaheuristic handles the goals and other factors either as a hard or as a soft target. The hard targets indicate that the solution may not include any violations for those targets. The soft targets indicate that violations may exist, but they should be minimized. A soft target is assigned a fixed weight 1, 2 or 3 (most important) according to our practical experience and recommendations (see [39]). The PEASTP metaheuristic uses a penalty method, which assigns positive weights to the hard and soft targets and sums the violation scores to get a single value to be optimized (see [35]).

The optimization task of the employee-based shift generation described in Sect. 2 can be compressed and simplified to the following three main goals. The shift structure is generated separately for each day.

A1. The shifts should match the predicted workload as accurately as possible
A2. The minimum and maximum length of the shifts must be respected
A3. The start and end times of the shifts must be respected.

We generated three different shift structures for the contact center based on the length of the shifts and the average shift length (see Table 3). The total number of required hours to cover the demand varies from 10617 to 12456 and the number of shifts from 978 to 1557. The average shift length varies between 480 and 651 min. Notably, the number of evening shifts (shift structure B) is greatly reduced when longer shifts are preferred (shift structure C). Figure 3 shows an example how well the optimized shift structure covers the predicted workload.

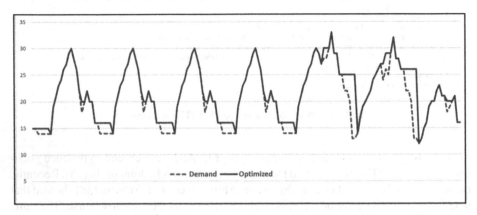

Fig. 3. The shift structure covering a one-week demand using B settings (see Table 3)

Next, we rostered the staff using these three different shift structures. The number of employees needed to roster all shifts varies between 113 (longest shifts) and 149 (shortest shifts). The optimization task of the staff rostering described in Sect. 2 can be compressed and simplified to the following three main goals. The length of the planning horizon is three weeks.

B1. A minimum number of employees with particular competences must be guaranteed for each shift.
B2. An employee's total working time should be within the given limits.
B3. The twelve stress and risk factors and the recommendations published by the Finnish Institute of Occupational Health should be considered.

Note that this study does not include the employees' requests due to the lack of data and the practical difficulty in gathering it for the different shift structures.

Table 3 shows the effects of the three different shift structures on the possibility to generate acceptable staff rosters. The five most evident observations and findings are the following:

1) Less employees are needed when the average shift length increases.
2) The demand is significantly easier to cover with longer shifts.
3) The stress and risk factors and the recommendations of FIOH are better considered with longer shifts.
4) Less evening shifts are needed with longer shifts.
5) Shift structure has no effect on the number of night shifts.

As much as 32% more employees are needed when comparing shift structures A and C. This is due to the 17% increase in the total number of required hours to cover the

Table 3. The effects of different shift structures on staff rosters.

SHIFT STRUCTURE	A	B	C
Main targets			
C2 and C3 (optimized shift coverage)	Working time in shifts must match demand		
V3 (min and max shift length)	8 – 8	6 – 12	6 - 12
V4 (average shift length)	8	10	max
P1 (shift start time)	Shifts cannot start between 22:01 and 06:59		
P2 (shift end time)	Night shifts must end before 08:01		
Characteristics of the generated shifts			
Total number of required hours	12456	10863	10617
Average shift length	480	525	651
Number of shifts	1557	1242	978
Nbr of morning, evening and night shifts	597, 570, 390	453, 399, 390	453, 142, 383
STAFF ROSTERING SOLUTION	**A**	**B**	**C**
Number of employees needed	149	125	113
Number of hard violations	55	36	19
Number of soft violations	566	116	47
0. Total working time compared to target time			
A. Within ±4 hours (S1)	200	7	1
1. Length of the workday			
A. Longest work period ≤ 40h (S3)	0	1	5
C. Consecutive working days 2-4 (H)	0	0	0
D. Single working days minimized (S1)	143	31	27
2. Timing of the working hours			
A. Consecutive evening shifts 0-4 (S1)	0	0	0
B. Consecutive night shifts 0 or 2-4 (S3)	12	10	1
C. Single night shifts not allowed (H)	0	0	0
D. Number of night shifts 0 or 2-4 (S3)	37	1	0
3. Recovery and rest times			
A. Rest times below 11h 0-1 (S1)	0	1	0
B. Rest time after a night shift ≥ 48h (S3)	10	6	0
C. Longest rest period ≥ 35h (H)	0	0	0
4. Reconciling work and family life			
A. Free weekends 1-3 (H)	55	36	19
B. Single days-off minimized (S1)	164	59	13

demand, and also partly due to the 60% increase in number of shifts. This, of course, would increase the payload. Even if it could be possible to hire more employees, adopting eight-hour shifts would increase the stress and risk factors to heavy and overload.

On the other hand, versatile shift lengths would enable employees to better target their shift requests, i.e. reconciling their work and family life. For example, an employee might want to have a shorter shift either in the morning or in the evening. However, Table 3 shows that with longer shifts the optimized rosters include more free weekends and less single days-offs.

According to FIOH, the ideal maximum shift length is ten hours. Table 3 shows that when the average shift length is ten hours (shift structure B), acceptable staff rosters can be generated. However, 11% more employees are still needed compared to shift structure C. Furthermore, the number of soft violations increases 250%.

An important observation not evident from the computational results is the effect of having less number of evening shifts. The adequate rest times between the shifts is then easier to realize. On the other hand, night shifts are the least preferred among the employees. The presented analysis does not provide a solution to this challenge.

5 Conclusions

To the best of our knowledge, this was the first encounter to examine the effect of shift structures on staff rostering optimization. The experiment was not too comprehensive, including only a single instance from a Finnish contact center. However, it underlines well the significant impact of shift structure on e.g. the number of required employees. In practice shift structure is often highly constrained by legislative or practical human considerations, greatly affecting efficiency. For example, the extreme demand peaks often encountered in airports would require unreasonably short shifts to fulfill without major overstaffing.

It was found that increasing the average shift length from 8 to 10 h yielded significant improvements in both shift structure and staff rosters. In the 4-year registry study by Vedaa et al. [40], long working hours was found to be associated with less sick leave days. The restorative effects of extra days off with long working hours were discussed as possible explanations to this relationship. In addition, in the shift work study considering industry workers in large Finnish companies [41], the employees were found to sleep and feel better when the length of the working day was twelve hours instead of eight hours. An interesting idea reflecting the shift structure is to introduce an adaptive rest time, i.e. the number of rest hours could depend on the number of consecutive hours worked.

Some people can be characterized as early and some as late chronotypes [42]. Late chronotypes' circadian schedules are shifted later by genetics. Early risers rule our society clock. Unfortunately, late sleepers cannot easily control their behavior. If we fight our chronotypes, our health may suffer. Individual factors also vary with age. With increasing age, tolerance of shift work usually decreases and the chronotype changes towards early [43]. Some individuals are more influenced by these effects than others are. Therefore, it is extremely important that employees have control over their working hours. This buffers the negative impact of shift work on their health and work-life balance. High levels of work time control have been shown to positively influence mental health outcomes such as affective well-being and perceived stress [44].

Every organization wants to keep its most valuable employees happy and engaged. Better staff rostering management with increased flexibility will improve employee satisfaction and retention [45]. An employee retention plan will help an organization to reduce employee turnover and to keep the top talents at the organization. Successful organizations realize that having an effective employee retention plan will help them sustain their leadership and growth in the marketplace [46]. Furthermore, better staff rostering management has a role in extending working life.

In summary, the study showed that the generated shift structures have a significant effect on optimizing the staff rosters. The study also showed that we should allow longer working days, because they imply better consideration of the stress and risk factors introduced by the Finnish Institute of Occupational Health. The PEASTP metaheuristic, a computational intelligence framework, justified the findings.

References

1. Musliu, N., Schaerf, A., Slany, W.: Local search for shift design. Eur. J. Oper. Res. **153**(1), 51–64 (2004)
2. Di Gaspero, L., Gärtner, J., Kortsarz, G., Musliu, N., Schaerf, A., Slany, W.: The minimum shift design problem. Ann. Oper. Res. **155**, 79–105 (2007)
3. Kyngäs, N., Goossens, D., Nurmi, K., Kyngäs, J.: Optimizing the unlimited shift generation problem. In: Di Chio, C., et al. (eds.) Applications of Evolutionary Computation. Lecture Notes in Computer Science, vol. 7248, pp.508–518. Springer, Berlin, Heidelberg (2012). https://doi.org/10.1007/978-3-642-29178-4_51
4. Kyngäs, N., Nurmi, K., Kyngäs, J.: Solving the person-based multitask shift generation problem with breaks. In: Proceedings of the 5th International Conference on Modeling, Simulation and Applied Optimization, pp. 1–8 (2013)
5. Dowling, D., Krishnamoorthy, M., Mackenzie, H., Sier, H.: Staff rostering at a large international airport. Ann. Oper. Res. **72**, 125–147 (1997)
6. Valls, V., Perez, A., Quintanilla, S.: A graph colouring model for assigning a heterogenous workforce to a given schedule. Eur. J. Oper. Res. **90**, 285–302 (1996)
7. Krishnamoorthy, M., Ernst, A.T.: The personnel task scheduling problem. Optim. Methods Appl., 343–367 (2001)
8. Krishnamoorthy, M., Ernst, A.T., Baatar, D.: Algorithms for large scale shift minimisation personnel task scheduling problems. Eur. J. Oper. Res. **219**(1), 34–48 (2012)
9. Jansen, K.: An approximation algorithm for the license and shift class design problem. Eur. J. Oper. Res. **73**, 127–131 (1994)
10. Kroon, L.G., Salomon, M., Van Wassenhove, L.N.: Exact and approximation algorithms for the tactical fixed interval scheduling problem. Oper. Res. **45**(4), 624–638 (1997)
11. Kolen, A.W.J., Lenstra, J.K., Papadimitriou, C.H., Spieksma, F.C.R.: Interval scheduling: a survey. Nav. Res. Logist. **54**(5), 530–543 (2007)
12. Nurmi, K., Kyngäs, N., Kyngäs, J.: Workforce optimization: the general task-based shift generation problem. IAENG Int. J. Appl. Math. **49**(4), 393–400 (2019)
13. Kyngäs, N., Nurmi, K.: The extended shift minimization personnel task scheduling problem. In: Position and Communication Papers of the 16th Conference on Computer Science and Intelligence Systems, vol. 26, pp. 65–74 (2021)
14. Garey, M.R., Johnson, D.S.: Computers and intractability: a guide to the theory of NP-completeness. Freeman (1979)
15. Tien, J., Kamiyama, A.: On manpower scheduling algorithms. SIAM Rev. **24**(3), 275–287 (1982)
16. Lau, H.C.: On the complexity of manpower shift scheduling. Comput. Oper. Res. **23**(1), 93–102 (1996)
17. Marx, D.: Graph coloring problems and their applications in scheduling. Periodica Polytechnica Ser. El. Eng. **48**, 5–10 (2004)
18. Bruecker, P., Qu, R., Burke, E.: Personnel scheduling: models and complexity. Eur. J. Oper. Res. **210**(3), 467–473 (2011)

19. Jackson,W.K., Havens, W.S., Dollard, H.: Staff scheduling: a simple approach that worked. Technical Report CMPT97-23, School of Computing Science, Simon Fraser University, Canada (1997)
20. Lapegue, T., Bellenguez-Morineau, O., Prot, D.: A constraint-based approach for the shift design personnel task scheduling problem with equity. Comput. Oper. Res. **40**(10), 2450–2465 (2013)
21. Prot, D., Lapgue, T., Bellenguez-Morineau, O.: A two-base method for the shift design and personnel task scheduling problem with equity objective. Int. J. Prod. Res. **53**(24), 7286–7298 (2015)
22. Smet, P., Ernst, A.T., Vanden Berghe, G.: Heuristic decomposition approaches for an integrated task scheduling and personnel rostering problem. Comput. Oper. Res. **76**, 60–72 (2016)
23. The Finnish Institute of Occupational Health, "Recommendations for shift work". https://www.ttl.fi/teemat/tyohyvinvointi-ja-tyokyky/tyoaika/vuorotyo/tyoaikojen-kuormittavuuden-arviointi. Accessed 17 Mar 2023. (in Finnish)
24. Karhula, K., et al.: Are changes in objective working hour characteristics associated with changes in work-life conflict among hospital employees working shifts? A 7-year follow-up. Occup. Environ. Med. **75**(6), 407–411 (2018)
25. Karhula, K., Hakola, T., Koskinen, A., Ojajärvi, A., Kivimäki, M., Härmä, M.: Permanent night workers' sleep and psychosocial factors in hospital work. a comparison to day and shift work. Chronobiol. Int. **35**(6), 785–794 (2018)
26. Nurmi, K., Kyngäs, J., Kyngäs, N.: Staff rostering optimization: ideal recommendations vs. real-world computing challenges. In: Arai, K. (ed.) Intelligent Computing. Lecture Notes in Networks and Systems, vol. 283, pp. 274–291. Springer, Cham (2022). https://doi.org/10.1007/978-3-030-80119-9_15
27. Nurmi, K., Kyngäs, J., Kyngäs, N.: Synthesis of employer and employee satisfaction - case nurse rostering in a finnish hospital. J. Adv. Inf. Technol. **7**(2), 97–104 (2016)
28. Kirkpatrick, S., Gelatt, C.D., Vecchi, M.P.: Optimization by simulated annealing. Science **220**, 671–680 (1983)
29. Glover, F.: Future paths for integer programming and links to artificial intelligence. Comput. Oper. Res. **13**(5), 533–549 (1986)
30. Goldberg, D.: Genetic Algorithms in Search, Optimization and Machine Learning. Addison Wesley, USA (1989)
31. Mladenovic, N., Hansen, P.: Variable neighborhood search. Comput. Oper. Res. **24**(11), 1097–1100 (1997)
32. Dees, W.A., Smith II, R.: Performance of interconnection rip-up and reroute strategies. In: 18th Design Automation Conference, pp. 382–390 (1981)
33. Schrimpf, G., Schneider, K., Stamm-Wilbrandt, H., Dueck, W.: Record breaking optimization results using the ruin and recreate principle. J. Comput. Phys. **159**, 139–171 (2000)
34. Glover, F.: New ejection chain and alternating path methods for traveling salesman problems. In: Computer Science and Operations Research: New Developments in Their Interfaces, pp. 449–509 (1992)
35. Kyngäs, N., Nurmi, K., Kyngäs, J.: Crucial Components of the PEAST algorithm in solving real-world scheduling problems. LNSE, 230–236 (2013). https://doi.org/10.7763/LNSE.2013.V1.51
36. Kyngäs, N., Nurmi, K., Kyngäs, J.: Workforce scheduling using the PEAST algorithm. In: Ao, S.-I. (ed.) IAENG Transactions on Engineering Technologies. Lecture Notes in Electrical Engineering, vol. 275, pp. 359–372. Springer, USA (2014). https://doi.org/10.1007/978-94-007-7684-5_25

37. Nurmi, K., Kyngäs, J., Järvelä, A.I.: Ten-year evolution and the experiments in scheduling a major ice hockey league. In: Daniel Hak (ed.) An in Depth Guide to Sports, Nova Science Publishers, pp 169–207 (2018)
38. Kyngäs, N., Nurmi, K., Goossens, D.: The general task-based shift generation problem: formulation and benchmarks. In: Proceedings of the 9th Multidisciplinary International Scheduling Conference: Theory and Applications (MISTA), Ningbo, China (2019)
39. Nurmi, K., Kyngäs, J., Kyngäs, N.: Practical recommendations for staff rostering justified by real-world optimization. In: Proceedings of the 11th International Conference on Health and Social Care Information Systems and Technologies (HCist) (2022)
40. Vedaa, Ø., et al.: Long working hours are inversely related to sick leave in the following 3 months: a 4-year registry study. Int. Arch. Occup. Environ. Health 92, 457–466 (2019)
41. Karhula, K., et al.: 12 tunnin vuorojärjestelmien turvallinen ja työhyvinvointia edistävä toteuttaminen teollisuudessa. Report 114114, Finnish Institute of Occupational Health (2016). (in Finnish)
42. Roenneberg, T., et al.: Epidemiology of the human circadian clock. Sleep Med. Rev. 11(6), 429–438 (2007)
43. Roenneberg, T., et al.: A marker for the end of adolescence. Current Biol. 14(24), R1038–R1039 (2004)
44. Nijp, H.H., Beckers, D.G., Geurts, S.A.: Systematic review on the association between employee worktime control and work-non-work balance, health and well-being, and job-related outcomes. Scand. J. Work Environ. Health 38, 299–313 (2012)
45. Wittmer, J., Shepard, A.K., James, E., Martin, J.E.: Schedule preferences, congruence and employee outcomes in unionized shift workers. Am. J. Bus. 30(1), 92–110 (2015)
46. Sumathi, N., Parimala, S.: Impact of HR benefits on employee's retention in surface transportation firms operating in Tamil Nadu. Int. J. Asian Dev. Stud. 28(6), 483–487 (2015)

Consumers Financial Distress: Prediction and Prescription Using Machine Learning

Hendrik de Waal[1], Serge Nyawa[2](✉) (iD), and Samuel Fosso Wamba[2] (iD)

[1] Johannesburg Business School, 69 Kingsway Ave, Auckland Park, Johannesburg 2092,
South Africa
hendrikdewaal@capitecbank.co.za
[2] Department of Information, Operations and Management Sciences, TBS Business School, 1
Place Alphonse Jourdain, 31068 Toulouse, France
{s.nyawa,s.fosso-wamba}@tbs-education.fr

Abstract. This paper shows how transactional bank account data can be used to predict and to prevent financial distress in consumers. Machine learning methods were used to understand what are the most significant transactional behaviours that cause financial distress. We show that Gradian Boosting outperforms the other machine learning models when predicting the financial distress of a consumer. We also obtain that, Fees, Uncategorised transactions, Other Income, Transfer, Groceries, and Interest paid were sub-categories of transactions which highly impacted the risk to be financially distressed. The study also proposes prescriptions that can be communicated to the client to help the individual make better financial decisions and improve their financial wellbeing by not entering a state of financial distress. This research used data from a major south African bank and the study was limited to credit card clients.

Keywords: Machine Learning · Classification · Deep Learning · Consumer Financial Distress · Explainability

1 Introduction

Usually defined as a state in which an entity is unable to meet its financial responsibilities because of a lack of income, financial distress is getting a lot of attention since the onset of the global pandemic. Living with financial distress can have a range of negative effects on an individual's physical, mental, and emotional well-being. Some of the effects of financial distress may include physical health problems, mental health problems, relationship problems, decreased job performance, poor decision-making and resulting reduced quality of life. (Sturgeon et al. 2016; Guan et al., 2022; Xiao and Kim, 2022).

Many banks and credit unions have made financial health and resiliency a top priority to assist their customers in overcoming financial uncertainty. This attention is also the result of increased oversight, from financial services industry regulators, to prevent customers from experiencing financial distress. Analysis into understanding what causes financial distress and developing interventions to prevent financial distress in individuals will be a major driver in improving general wellbeing.

© The Author(s), under exclusive license to Springer Nature Switzerland AG 2024
H. Moosaei et al. (Eds.): DIS 2023, LNCS 14321, pp. 218–231, 2024.
https://doi.org/10.1007/978-3-031-50320-7_16

The capacity to predict financial distress, to detect its early signs and to understand how this distress is related to consumer credit use is therefore essential. From a business perspective, it can help banks to detect financial difficulties earlier, to understand its process, to prevent the occurrence of insolvency, and to ensure a sustainable and profitable enterprise. Financial distress in clients means that clients are in arrears on credit products which can ultimately result in clients defaulting causing increased bad debts for financial service providers (Zhang, Wang and Liu, 2023). Any analysis that can enhance the understanding of the drivers of financial distress can inform strategies to alleviate default levels and lessen credit losses for the business. Similarly any initiatives or interventions that will assist consumers to make better financial decisions, that will prevent them from defaulting on credit agreements, will have a direct impact on the profitability of financial service providers (Markov, Seleznyova and Lapshin, 2022). Research into financial distress can therefore not only advance the understanding of human financial behaviour but can also inform on changing economic factors that influence the financial wellbeing of individuals and businesses.

Research related to consumer's financial distress is organized into two main directions. The first aims to understand the causes of financial distress, by looking how bad financial behaviour or external factors like the economy or job-loss can generate financial difficulties. The second investigates the impact that financial distress in consumers has on the financial services providers. This paper focus on understanding the behavioural factors that contribute to financial distress in consumers and proposes (prescribes) interventions that can assist consumers to make good financial decisions that will improve their ability to live with less financial distress and so doing improve their general wellbeing and the wellbeing of those around them. Our research questions are as follows:

1. How can machine learning (ML) efficiently predict consumer's financial distress using transactional data?
2. Based on an explainable ML model, which factors drive consumer's financial distress?
3. What is the profile of a typical banking client in financial distress?
4. What are the main remedies for financial distress?

Based on transactional data, this research offers detailed ML models to predict consumer's financial distress, such as: Random Forest, Decision tree, Gradient boost (Gboost), Logistic regression, Naive Bayes, Support vector classifier, and Artificial neural networks (ANNs) were the seven ML models we took into consideration. Our strategy opens possibilities for cutting-edge study in financial distress. To do this, we used consumer transactional data from a major south African bank and the study was limited to credit card users.

We contribute to the literature of individual's financial distress in two ways. We started by addressing the lack of studies using ML models to predict financial distress. For data preparation, model training, validation, and evaluation, we have developed transparent steps. We made sure the techniques were compared using the same dataset and in the same situation. This approach therefore guarantees the reliability and repeatability of our results.

The remainder of the paper is structured as follows. The related literature is presented in Sect. 2. In Sect. 3, we then go over the methodology, and Sect. 4 presents the findings. A discussion is included in Sect. 5, and Sect. 6 concludes.

2 Related Literature

The application of advanced business analytics is an essential part of modern banking. Business analytics can broadly be divided into 3 categories namely descriptive analytics, predictive analytics and prescriptive analytics (Schniederjans, Schniederjans and Starkey, 2014; Sharma et al., 2022).

Descriptive analytics is aimed at understanding the factors that influence an outcome (explain why things happen). Predictive analytics is aimed at predicting an outcome (what will happen). Prescriptive analytics focusses on prescribing an action that will influence an outcome (how do we influence what will happen). In this study we aim to apply all three techniques by predicting which clients will go into financial distress, explaining what the main drivers of financial stress are and proposing interventions that can prevent the client form going into financial distress.

Of all the risks that must be managed by financial institutions, credit risk is the most significant as it has a direct impact on the balance sheet and income statement in the form of bad debt (Markov, Seleznyova and Lapshin, 2022). Credit risk management therefore impacts product development, pricing, and marketing strategy.

Lenders use advanced analytical methods to develop scorecards to rank credit risk in individuals (Onay and Öztürk, 2018). Lenders will use a scorecard to determine who they can lend to or not based on their risk appetite. Scorecards can also be used to determine lending rates as the higher the score (the lower the risk) the lower the lending rate needed to offset possible bad debts conversely the lower the score (the higher the risk) the higher the lending rate needed to offset bad debts (Markov, Seleznyova and Lapshin, 2022).

Broadly speaking there are two kinds of score cards namely application and behavioural score cards (Li and Zhong, 2012). Application score cards are designed to rank the risk of new clients based on information and data gathered at the point of application. Behavioural score cards, on the other hand, are designed to develop an ongoing view of the risk associated with an existing client based on the historic data of the client behaviour on record. While application risk prediction is used to sell credit to new clients, behavioural risk prediction is used to sell new products to existing clients (e.g. credit card limit increases).

While this study is applied on existing credit card clients the aim is not to develop a behavioural scorecard that can be used to advance new credit. The objective of this study is to use historical transactional data to predict which existing credit card clients will become financially distressed in the future, to determine the main drivers of financial distress in these clients and to prescribe interventions that will prevent clients from going into financial distress. The prevention of financial distress in consumers will in turn lead to improved wellbeing of the individual as well as improved profits for the lending institution.

As lenders develop better methods of identifying financial distress drivers and effective interventions to prevent financial distress more clients will not default on credit products allowing banks and lenders to provide financial services to a wider part of the population and so doing advance financial inclusion into sectors of the society that have previously been excluded due to perceived high risk.

3 Data and Methodology

3.1 Data Preparation

The following context diagram shows the data structure that was used in the analysis. The green is pre-existing processes within the bank while the blue was developed for this analysis. Step 1 was carried out first to extract the credit performance data of a random set of credit card clients that were active during the analysis period. Step 2 was carried out to extract categorised transaction data on the same set of clients selected in step 1. Step 1 created the target data for the prediction modelling while step 2 created the input data (Fig. 1).

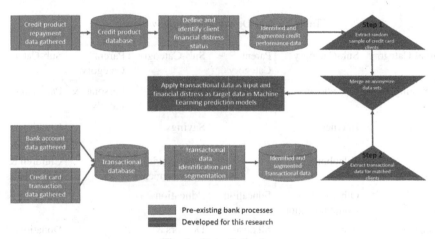

Fig. 1. Data Collection

Input Variables. The following section summarises the transactional features that were used in the study. Raw banking data is already categorised into transaction types by the banks own internal processes. This categorisation has been developed over time in an environment where clients can interact with the categorisation and make corrections in real time. The categorisation is done by a voting-based Machine Learning process that continuously learns from the client inputs (votes). 93% of all transactions are classified. New categories are continuously added to maintain this high level of client validated categorisation.

The transactional data includes transactions that result in money flowing into the client account and transactions that result in money flowing out of the client account.

The following is a list of the inflows identified in the transactional data (Table 1).

The list of the outflows identified in the transactional data is given below (Table 2).

Table 1. Data Description, inflow of money

Money in categories			
Parent Category	Sub-Category	Parent Category	Sub-Category
Salary	Salary	Pension	Pension
	Commission	Allowance	Allowance
	Bonus	Cash Deposit	Cash Deposit
	Overtime	Interest	Interest
Other Income	Other Income	UIF Payments	Unemployment Insurance Fund Payments
Rental Income	Rental Income	Banking Rewards Benefit	Banking Rewards
Investment Income	Investment Income	Child Support	Child Support
Loans	Loans		

Table 2. Data Description, outflow money

Parent Category	Sub-Cateogry	Parent Category	Sub-Cateogry	Parent Category	Sub-Cateogry
Communication	Cell phone	Savings & Investments	Investments	Personal & Family	Personal Care
	Internet		Savings		Clothing & Shoes
	Telephone		Other Saving & Investments		Children & Dependants
	Other Communication	Education	Education		Gifts
Transport	Fuel	Medical	Doctors & Therapists		Donations
	Public Transport		Pharmacy		Pets
	Vehicle Maintenance		Medical Aid		Sport & Hobbies
	Licence		Other Medical		Holiday
	Tolls	Insurance	Vehicle Insurance		Legal Fees
	Parking		Funeral cover		Tax
	Vehicle Tracking		Home insurance		Gadgets

(continued)

Table 2. (*continued*)

Parent Category	Sub-Cateogry	Parent Category	Sub-Cateogry	Parent Category	Sub-Cateogry
	Other Transport		Life Insurance		Online Store
Entertainment	Digital Subscriptions		Other Insurance		Other Personal & Family
	Going Out	Loans & Accounts	Loan Payments	Household	Home Improvements
	Alcohol		Home Loan Payments		Home Maintenance
	Tobacco		Vehicle Payments		Furniture & Appliances
	Activities		Credit Card Payments		Housekeeping
	Software/Games		Store Account Payments		Rent
	Movies		Other Loans & Accounts		Levies
	Betting/Lottery	Food	Restaurants		Security
	Other Entertainment		Takeaways		Electricity
			Groceries		Water
			Other Food		Gas
		Cash	Cash Withdrawal		Municipal Bill
					Garden
					Other Household

The Target Variable. In our analysis, we define financial distress as being behind with at least one payment of a credit product with the bank (one month or more in arrears). The analysis is limited to credit card clients, but financial distress is defined as being in arrears on any credit product with the bank.

In a data preparation stage, the dataset is collected, cleaned, preprocessed, and transformed into a format suitable for training. This includes tasks such as anomaly detection, normalization, and feature engineering. In the resulting dataset, each sub-category is a feature, and its value for a given individual is the number of times this sub-category of transactions has been observed for that specific individual. The following figure illustrate the initial structure of the dataset and the resulting after the preparation step (Fig. 2).

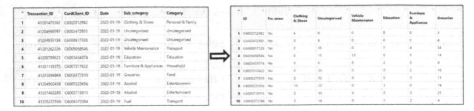

Fig. 2. Data Preparation

In the training stage, the prepared data is used to train a machine learning model using an appropriate algorithm. The training process involves adjusting the model parameters to minimize a specific loss function that measures the difference between the predicted and actual outputs. In the validation stage, a portion of the data that was not used in training is used to evaluate the performance of the trained model and to choose the optimal tuning parameters. In the testing stage, a separate set of data is used to evaluate the final performance of the model. This provides a more accurate estimate of how the model will perform on new, unseen data in the real world.

We created our training and testing sets by dividing the initial sample into two groups with proportions of 80% and 20% using our prepared dataset. Then, models were trained using a 5-fold cross-validation, and before the algorithms were finally put into practice, the best hyperparameters for each algorithm were determined.

As shown in the findings section, our cross-validation exercise produced a collection of hyperparameters for each ML algorithm.

3.2 Methodology

Machine Learning (ML) focuses on creating algorithms that can learn from data and make prediction about it. It is focused on developing models that, without being specifically programmed to do so, can automatically improve their performance on a task. Prediction by ML is achieved by analyzing historical data to identify patterns and relationships between variables, and then using these insights to train a machine learning model.

Seven main steps are often used to perform a prediction using ML: i) *data collection*: the first stage in the machine learning process is to gather the pertinent data. The information used must be indicative of the issue you are attempting to address; ii) *data preparation*: consists of cleaning, preprocessing, and converting the data into a structure suitable for the machine learning model's training; the handling of missing numbers, feature scaling, and feature selection fall under this category; iii) *model Selection*: to make predictions based on the data, you must choose a suitable machine learning model; the problem you're attempting to solve, the kind of data you have, and the performance metrics you want to optimize will all influence the model you choose; iv) *model training*: a suitable method is then used to train the chosen model on the data; the model gains the ability to recognize patterns and connections in the data that can be used to forecast the outcome of new data inputs during training; v) *model evaluation*: after the model has been trained, its performance must be assessed on a different set of data; the accuracy, precision, recall, and other performance measures are measured in this process; vi) *model optimization*: the model is adjusted through the hyperparameters or choosing an alternative algorithm based on the evaluation results; vii) *predictions*: on fresh, unexplored data, forecasts are made using the trained and improved algorithm. The quality of the training data, the model selected, and the algorithm's success will all affect how accurate the predictions are.

By providing a concise and understandable explanation of our ML algorithms, we hope to help boost users' trust in ML models. Business users are growing more and more curious to know how these algorithms work and how they produce knowledge about domain connections into data.

Decision Tree. The main goal of a decision tree is to represent choices and potential outcomes in the form of a tree. The tree is made up of nodes that stand in for decision-making points and branches that symbolize potential outcomes. The root node, which serves as the algorithm's starting point, is a single node that represents the complete dataset. The algorithm chooses a feature that it thinks is most informative for the current classification job at each decision point. The dataset is then divided according to the value of the feature, with a branch being created for each potential value. For each new branch, this process is repeated iteratively until either a stopping criterion is reached or all data points in a particular branch have the same classification label (see, e.g., Breiman et al. (1984), de Ville et al. (2014)).

Random Forest. The ensemble learning method called random forest combines various decision trees to produce more accurate forecasts. Random sampling is used to generate several data sets known as "bootstrap samples" using training data. For each bootstrap sample, a decision tree is created by selecting the best feature to split the data on at each node. The goal is to create decision trees that have high accuracy and low correlation with each other. The collection of decision trees produced in previous steps is then assembled. For classification issues, this is accomplished by taking the majority (see, e.g., Liaw et al. (2002), Ishwaran et al. (2014)).

Adaptative Boosting and Gradient Boosting. Adaptive Boosting and Gradient Boosting combine multiple weak learners to create a strong learner that is better at making predictions. In the beginning of AdaBoost, each instance in the training data is given equal importance. A weak learner, such as a decision tree with a single split, is trained on the data, and the instances that were misclassified are given more weight. This process is repeated multiple times, with each iteration placing more emphasis on the misclassified instances, until a predefined number of weak learners is reached. Finally, the weak learners are combined to create a strong learner that can make accurate predictions. Gradient Boosting also works by combining multiple weak learners, but by fitting each weak learner to the residuals of the previous weak learner. In the beginning, a weak learner, such as a decision tree, is trained on the data. Then, the errors made by this weak learner are calculated, and another weak learner is trained to predict these errors. The two weak learners are combined, and the process is repeated multiple times, with each iteration improving the prediction accuracy. Finally, the weak learners are combined to create a strong learner that can make accurate predictions (see, e.g., Hastie et al. (2009), Friedman et al. (2001)).

Naïve Bayes. The foundation of Nave Bayes is the Bayes' theorem, a mathematical formula that estimates the likelihood of an event based on information of circumstances that might be connected to it in the past. In Naive Bayes, it is assumed that the presence or absence of a particular feature is independent of the presence or absence of any other feature. As a result, probability computations are streamlined, and the algorithm becomes quicker and more effective. For a given input, the algorithm determines the probabilities of each possible class, and then it chooses the class with the greatest probability as the predicted output (see, e.g., Mehrotra (2015), Domingos et al. (1997)).

Support Vector Classifier (SVC). The SVC operates by establishing a decision boundary in the feature space that divides two classes of data points. Finding the hyperplane that optimizes the margin between the two classes results in the creation of the boundary. The margin, also referred to as support vectors, is the separation between the decision boundary and the nearest data values from each class. The goal of the convex optimization problem that generates the hyperplane is to increase the margin between the two classes within certain bounds. Constraints guarantee that the data elements are properly classified and that they are situated on the right side of the decision boundary. When the data cannot be separated linearly, SVCs are especially useful. In such cases, the SVC can use kernel functions to map the original data into a higher-dimensional space where it becomes linearly separable.

Logistic Regression. The logistic regression model assumes that the relationship between the independent variables and the dependent variable is log-linear. That is, the log-odds of the dependent variable taking a value of 1 is a linear function of the independent variables. The log-odds is also known as the "logit" function, which maps a probability value between 0 and 1 to a value between negative infinity and infinity. The logistic regression model is estimated using maximum likelihood estimation, which involves finding the parameter values that maximize the likelihood of the observed data given the model. The parameters of the model are typically estimated using iterative numerical optimization algorithms.

Artificial Neural Networks (ANNs). Artificial neural networks are inspired by the structure and function of the human brain. ANNs are composed of interconnected nodes (also known as neurons) that process and transmit information. The structure of an ANN typically consists of an input layer, one or more hidden layers, and an output layer. Each node in a layer is connected to every node in the adjacent layer, and each connection has an associated weight. During training, the weights are adjusted so that the ANN can accurately map input data to output predictions. The activation function of each node determines its output value based on the weighted sum of its inputs. There are various activation functions, such as the sigmoid function, which is commonly used in binary classification tasks, or the ReLU function, which is commonly used in image classification tasks.

4   Findings

We used a set of potential hyperparameters through a 5-fold cross validation for each machine learning method, as indicated in the following Table 3.

Table 3. Hyperparameters used for each machine learning algorithm

ML Methods	Hyperparameters	Values
Neural Networks	Network architecture (hidden_layer_sizes)	100, (33,16), **(100, 50)**, (100,50,50)
	Activation function (activation)	**relu**, logistic
	Learning rate (learning_rate_init)	**0.001**, 0.003, 0.005, 0.1
	Optimizer (solver)	adam, **sgd**
Random Forest	Max depth of decision tree (max_depth)	**8**, 16, **32**, 64
	Number of decision trees (n_estimators)	**200**, 300, 500, 700, 1000
	criterion	**gini**, entropy
	bootstrap	True, **False**
Gradient Boosting	Decision tree max depth (max_depth)	**3**, 8, 16, 32, 64
	Number of estimators (n_estimators)	**150**, 200, 300, 500, 700, 1000
	Learning rate (learning_rate)	0.01, **0.1**, 0.3, 0.5
	Loss function (loss)	deviance, **exponential**
Support Vector Classifier	Kernel type (kernel)	**rbf**, sigmoid
	kernel coefficient (gamma)	**scale**, auto
	Regularization parameter (C)	**1.0**, 2.0, 3.0
	Shrinking heuristic usage (shrinking)	True, **False**
	Degree polynomial kernel function (degree)	**3**, 4, 5
	Enabling probability estimates (probability)	True, **False**
Naive Bayes	laplace (controlling Laplace smoothing)	**0**, 1
	type	**raw**, "no raw"
Logistic Regression	Family	**binomial**, gaussian, Gamma, inverse.gaussian, poisson
	Method	**glm.fit**, model.frame
Decision Tree	Splitting index (split)	**information**, gini
	Complexity parameter (cp)	**0.01**; 0.001; 0.0001

(continued)

Table 3. (*continued*)

ML Methods	Hyperparameters	Values
	Smallest number of observations in the parent node (minsplit)	**2;** 5; 10; 15
	Smallest number of observations in a terminal node (minbucket)	**1;** 5; 10; 15
	Maximum depth (maxdepth)	5; 10; 20; **30**

We compare our machine learning algorithms based on the accuracy, since prediction is the main objective of this exercise. The table below shows that the Gboost and the Random Forest algorithms have the highest accuracy (Table 4).

Table 4. Comparison of the effectivity of the different Machine Learning analysis techniques.

	TPR	FNR	FPR	TNR	Accuracy	Precision	Recall	F1
Decision Tree	14771	2419	7174	7443	0.6984	0.6649	0.8593	0.7497
Naive Bayes	17187	3	14581	36	0.5415	0.9979	0.9998	0.9989
Random Forest	13411	3779	5003	9614	**0.7239**	0.5825	0.7802	0.6670
ANN	13314	3876	5164	9453	0.7158	0.5848	0.7745	0.6664
SVC	13314	3876	5164	9453	0.7158	0.5848	0.7745	0.6664
Gboost	13264	3926	4837	9780	**0.7245**	0.5756	0.7716	0.6593
Logistic Regression	2809	138	14381	14479	0.5435	0.1625	0.9532	0.2776

Then, the best model was selected to identify the most relevant features. Thus, the Gboost method was used to determine the most influential variables in predicting financial distress or financial health. The table below shows the most important variables identified by the Gboost analysis to predict if an individual will become financially distressed or remain financially healthy. While this set of variables are the most influential in the model, each of these variables can either contribute to financial distress or financial health and further analysis will be needed to determine exactly what role each plays. What is important though is that these variables represent the most influential factors that determine financial health in this group of credit card clients.

What will be of specific interest in further studies will be to understand the role that "Uncategorised" transactions and "Other Income" play in financial health. While the presence of uncategorised transactions is clearly predictive, the nature of these transactions cannot yet be identified automatically. Further research can be done to identify the transactions that constitute the "Uncategorised" variable in the current data set.

Table 5. Ranking of the most influential transaction variables.

Variable Importance (Gradian Boosting)	
Fees	100.0
Uncategorised	57.57
'Other Income'	13.77
Transfer	12.96
Groceries	7.459
Interest	4.221
Restaurants	0.718
Fuel	0.322
Takeaways	0.318
'Cash Deposit'	0.265
Pharmacy	0.252
Tolls	0.014
Rewards	0.0003
Gadgets	0.0002
Tobacco	0.0001

While many clients that have credit cards will earn "Salary" based income and be permanently employed other sources of income, represented by "Other Income" and "Cash Deposits", can provide extra financial support and security to individuals while they earn a salary from formal employment. These other sources of income could also enable an individual to survive financially in the event of being either temporarily or permanently unemployed.

Table 5 highlights the important role that factors like food expenses (groceries, takeaways, restaurants), transport expenses (Fuel, toll gates) and medical expenses play in the lives of consumers. Propper budgeting can play an important role in managing food and transport related expenses. Risk management tools like medical insurance and medical aid schemes can also be used to manage unforeseen medical expenses.

These findings can be used to develop prescriptive interventions that will support financial health and prevent financial distress. If individuals are predicted to be highly likely to become financially distressed in future the following interventions can be prescribed:

1. Communicate the importance of creating an extra source of income and provide information on how to start a small or informal business aimed at strengthening their financial resilience.
2. Communicate the importance of having a budget. They can be provided with a budget tool that can assist them to create a budget based on their actual income and affordable expenses. The tool should also assist individuals to track their income and spending

patterns against expectation. Many banks and neo-banks have integrated budget or Personal Financial Management tools available on their digital platforms that the client can be referred to. These tools help a client set up and stick to a budget and achieve financial goals.

3. Communicate the importance of planning for unforeseen medical expenses. Provide the individual with information of available medical insurance and medical aid schemes. Many financial service providers can also provide in house medical insurance products.

5   Limitations of This Study and Possible Enhancements for Future Studies

1. Future studies can enhance the analysis by using Machine Learning methods that not only identify the drivers that influence financial distress but also identify if the impact is a positive or negative effect on financial distress.
2. The current study used a 3-month outcome period. The study could be expanded to analyse whether predicting financial distress over a longer outcome period is more accurate.
3. The current study used the number of incidences of a transaction type as input variable. This could be enhanced by also using the value of a transaction type to predict financial distress.
4. The scope of the analysis can be expanded by including more than just credit card clients (e.g., Term Loans, Vehicle Finance and Home Loans). The type of credit used by the client could also be of interest.
5. This study can be enhanced by focusing more on the prescriptive measures that are proposed to assist clients to change behaviour and prevent financial distress.
6. The creation of compound features where the input variables are grouped or combined into new variables could also provide a more accurate prediction of financial distress.
7. This study shows that transactional banking data can be used to predict financial distress. This can be very meaningful in environments where there is no credit bureau data available. Future studies could apply a similar research method to predict financial distress using mobile wallet transactional data. This could be meaningful in environments where traditional transactional banking has been replaced by mobile wallets ecosystems.

6   Conclusion

The study shows that transactional bank account data and Machine Learning techniques can successfully be used to predict financial distress in credit card clients. Seven Machine Learning methods were used, with Gboost proving to be the most accurate. The study also shows that the predictions can be explained by identifying which transaction variables are the most influential to predict financial health. Based on the most influential variables identified by the analysis we have proposed three prescriptive interventions which will support financial health and prevent financial distress.

References

Breiman, L., Friedman, J., Stone, C., Olshen, R: Classification and Regression Trees. Chapman and Hall (1984)

de Ville, B., Watson, P.: Decision Trees for Business Intelligence and Data Mining: Using SAS Enterprise Miner. SAS Institute (2014)

Domingos, P., Pazzani, M.: On the optimality of the simple bayesian classifier under zero-one loss. Mach. Learn. **29**, 103–130 (1997)

Friedman, J.H.: Analytics in banking: time to realize the value | McKinsey (2017). https://www.mckinsey.com/industries/financial-services/our-insights/analytics-in-banking-time-to-realize-the-value. Accessed 2 Apr 2023

Guan, N., et al.: Financial stress and depression in adults: a systematic review. PLOS One **17**(2), e0264041 (2022). https://doi.org/10.1371/journal.pone.0264041

Li, X.-L., Zhong, Y.: An overview of personal credit scoring: techniques and future work (2012). https://doi.org/10.4236/ijis.2012.224024

Markov, A., Seleznyova, Z., Lapshin, V.: Credit scoring methods: latest trends and points to consider. J. Financ. Data Sci. **8**, 180–201 (2022). https://doi.org/10.1016/j.jfds.2022.07.002

Onay, C., Öztürk, E.: A review of credit scoring research in the age of big data. J. Financ. Regulation Compliance **26**(3), 382–405 (2018). https://doi.org/10.1108/JFRC-06-2017-0054

Schniederjans, M.J., Schniederjans, D.G., Starkey, C.M.: Business Analytics Principles, Concepts, and Applications: What, Why, and How. Pearson Education (2014)

Sharma, A.K., et al.: Analytics techniques: descriptive analytics, predictive analytics, and prescriptive analytics. In: Jeyanthi, P.M., Choudhury, T., Hack-Polay, D., Singh, T.P., Abujar, S. (eds.) Decision Intelligence Analytics and the Implementation of Strategic Business Management. EAI/Springer Innovations in Communication and Computing, pp. 1–14. Springer, Cham (2022). https://doi.org/10.1007/978-3-030-82763-2_1

Sturgeon, J.A., et al.: The psychosocial context of financial stress: Implications for inflammation and psychological health. Psychosom. Med. **78**(2), 134–143 (2016). https://doi.org/10.1097/PSY.0000000000000276

Team, M.: PFM solutions for banks, moneythor, 1 April 2023. https://www.moneythor.com/2021/04/01/pfm-solutions-for-banks/. Accessed 7 Aug 2023

Xiao, J.J., Kim, K.T.: The able worry more? Debt delinquency, financial capability, and financial stress. J. Family Econ. Issues **43**(1), 138–152 (2022). https://doi.org/10.1007/s10834-021-09767-3

Zhang, L., Wang, J. and Liu, Z.: What should lenders be more concerned about? Developing a profit-driven loan default prediction model. Expert Syst. Appl. **213**, 118938 (2023). https://doi.org/10.1016/j.eswa.2022.118938

Friedman, J.H. : Greedy function approximation: a gradient boosting machine. Ann. Stat. **29**(5), 1189–232 (2001). http://www.jstor.org/stable/2699986. Accessed 23 Oct 2023

Hastie, T., Rosset, S., Zhu, J., Zou, H.: Multi-class adaboost. statistics and its. Interface **2**(3), 349–360 (2009)

Ishwaran, H., Kogalur, U.B.: Random forests for survival, regression, and classification (RF-SRC). R News **2**(3), 18–22 (2014)

Liaw, A., Wiener, M.: Classification and Regression by random Forest. R News **2**(3), 18–22 (2002)

Mehrotra R.: Understanding naive bayes classifier: from theory to implementation. Int. J. Comput. Appl. (2015)

Optimizing Classroom Assignments to Reduce Non-essential Interactions Among University-Level Students During Pandemics

Mujahid N. Syed$^{(\boxtimes)}$ (iD)

Department of Industrial and Systems Engineering, Interdisciplinary Research Center
for Intelligent Secure Systems, King Fahd University of Petroleum and Minerals,
Dhahran 31261, Kingdom of Saudi Arabia
snumujahid@gmail.com, smujahid@kfupm.edu.sa

Abstract. Academic sectors involve students from multiple origins, multiple social circles, and different health conditions. During pandemics like COVID-19, a key challenge is to reduce non-essential interactions among students while they are at school/university. In this work, we focus on interactions that occur when university level students move between consecutive classes. Specifically, the interactions that arise during the movements of students between different university buildings via buses or cars. These types of movements are very critical during the pandemic because they are inevitable, and they occur simultaneously as well as periodically within small time window (10 to 15 min). Movement via buses is a hot spot for spreading viral diseases like COVID-19. Furthermore, high usage of cars/bikes during the 10 to 15 min class interval results in high traffic on campus roads, which in-turn leads to longer travel times of the buses (due to the road congestion). Nevertheless, having consecutive classes within walk-able range may reduce the above interactions. To sum, careful assignment of classrooms to courses reduces the above non-essential interactions. In this work, we present an mathematical modeling based approach that assigns classroom locations to courses such that the overall interactions are minimized. Specifically, we propose a novel mixed integer program (MIP) that minimizes the above interactions and incorporates behavior of the students. Numerical example is provided to showcase the implementation and effectiveness of the proposed MIP.

Keywords: Mixed Integer Program · Classroom Assignment Problem

1 Introduction

One of the challenging tasks during COVID-19 pandemic is the proper functioning of academic sectors. Indeed, non-academic work sectors are affected by the pandemic as well, however, operations with social distancing can be done

H. Moosaei et al. (Eds.): DIS 2023, LNCS 14321, pp. 232–241, 2024.
https://doi.org/10.1007/978-3-031-50320-7_17

effectively in the work sectors. Academic sectors involve students from multiple origins, multiple social networks, and different health conditions; meeting at a common place (the school/university). Adherence to the social distancing norms and guidelines among students may not be that effective when compared to the individuals from the work sectors. In addition to that, online academic operations may not be effective in monitoring progress of individual students, or in assessments. Thus, in-person academic operations are crucial for academic sectors, and they require more than social distancing guidelines to reduce the non-essential interactions among the students.

Our focus in this paper is on the university level students and their social distancing during pandemics. The most common event where the social distancing norms can be at risk in the university is during the movement of the students between classes. Inevitable interactions among the students that occur inside buses, during the movement between classes, can be reduced by:

- increasing the number of buses
- promoting movements via personal vehicles
- prolonging the time duration between consecutive classes
- optimizing assignment of classrooms locations to the courses

Typically, classroom locations in many public and private universities are decentralized decision making problems. They are trivially made by departments or colleges within a university. In this work, we propose the usage of mathematical model to develop a plan for academic sector administrators, which can reduce non-essential interactions among the students.

The rest of the paper is organized as follows: In Sect. 2, a short literature survey on the classroom assignment problem is presented. Section 3 depicts the novel mathematical model proposed in this work. Numerical experiment illustrating the effectiveness of the proposed approach is provided in Sect. 4. Finally, concluding remarks and future research directions are presented in Sect. 5.

2   Literature Note

Usage of mathematical model(s) in decision making is prevalent in the data driven era, where resources are scarce, and sustainability is of at-most importance. Assignment problem is a well know application area of operations research, that can be traced back to 1950's [7, 8, 10]. The aim of assignment problem is to assign one set of objects to another set of objects. Typically, the cost (or profit) of assignment is measured by linear or quadratic function. The objective of the assignment problem is then evaluated by minimizing the cost of assignment (or maximizing the profit of assignment). Our focus in the paper is on Classroom Assignment Problem (CAP). Frequently, CAP is solved right after timetable generation problem. These two problems together are part of a larger problem called timetabling problem [3, 14]. Since the timetabling problem is generally large-scale and difficult to solve, practically, time first rooms second strategy is used to decompose the timetabling problem into timetable generation and classroom assignment problems. Different versions of the CAP and the corresponding difficulties in solving the problem are highlighted in [4].

Typically, CAP attempts to find a mapping between set of course sections and set of rooms. If a linear cost measure is used, then the problem can be solved in polynomial time [4]. Various integer programming based CAP formulations are available in the literature (see [2,11]). Both exact and heuristics methods are developed for solving various instances of CAP [3,5].

In this work, the behavior of students (while waiting for the next class), and total interactions with other students (while traveling between the classrooms) is being incorporated in CAP. To the best of our knowledge, none of the earlier models considered the students' behavior and interaction in the CAP. We model the CAP as a Quadratic Assignment Problem (QAP). The QAP is as old as the assignment problem. However, unlike the assignment problem QAP is in general difficult to solve [1,9]. The quadratic objective function of proposed CAP attempts to measure the interactions and behavior of the students. Specifically, the behavior of the students is modeled via probabilities, which can be estimated via surveys or data collection. The objective of the proposed CAP is to minimize the total interactions. In the next section, the model is presented in detail.

3   Mathematical Model

The primary goal of our mathematical model is to capture the uncertainty in students' behavior while moving between the classes. In order to model the uncertain behavior, consider the illustration presented in Fig. 1. The arrows at the top indicate different time slots in each day. Furthermore, the vertical bars indicate list of courses that are scheduled during that time slot. Students after finishing a class, say $C_{1,1}$, can pick one of the multiple available options, depicted by arrows in Fig. 1. Furthermore, every individual student within an option may take different moves. For example consider a group of students who will be attending $C_{1,1}$ and $C_{3,1}$, where a student from the group after attending $C_{1,1}$ can have the following possible actions:

- may wait at the same location and move to the new location later,
- may move to the next location and wait at the new location,
- may move to the student housing at the end of T_1 and return later,
- may skip $C_{3,1}$ class,
- may skip rest of the classes altogether, etc.

This behavior of the student can be captured using the probabilities, as illustrated in Fig. 2, where a probability is assigned to each of the above actions. The above actions may not represent comprehensive set of actions; however, they are typical actions taken by the students. Furthermore, the complex behavior of the students can be approximated to probabilistic movements between two consecutive time periods. Moreover, the action selected by the student will influence the movement density of the later courses. In this work, we make a naive assumption that the student behavioral dynamics is independent among different time slots and days. That is, any option that the student select right after the class in time slot (T_1) will be independent from the option taken by the student after class in time slot T_2. Also, the option taken on a particular day at T_1 will be independent of the option taken on the same day of the week in any other week.

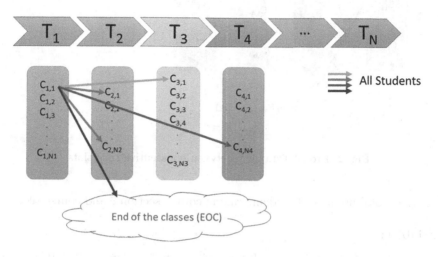

Fig. 1. Flow between time slots

In order to build the mathematical model, we are given with the information of students' course registration, as well as building/rooms that available on the campus. From the above information, following sets and parameters are created.

Sets:

- C: a set containing unique section numbers for all the courses that are offered in the university.
- H: a set containing unique dummy section numbers corresponding to student housing and/or cafeteria.
- T: a set containing unique time slots for all the courses that are offered in the university.
- E: a dummy section number corresponding to end of classes.
- $\overline{C}$: be defined as $C \cup H \cup E$.
- RM: a set containing unique room numbers available for the classrooms.
- O_t: set of courses that are offered during time period t, where $t \in T$.

Parameters:

- $A_{l,\bar{l}} = 1$ if location $l \in RM$ and location $\bar{l} \in RM$ belong to the same building (or they are at walk-able distance), 0 otherwise.
- $B_{c,l} = 1$ if course section $c \in CRN$ can be assigned to location $l \in RM$, 0 otherwise.

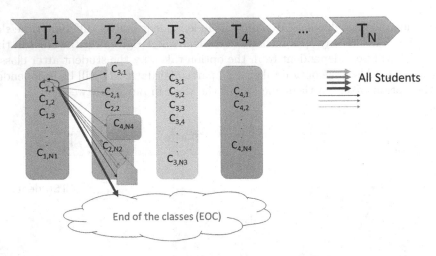

Fig. 2. Probabilistic flow between consecutive time slots

– $N_{c,\bar{c}}$ = total number of students taking course section $\bar{c}$ and course section c.

Variables:

– $x_{c,l} = 1$ if course section $c \in CRN$ is assigned to location $l \in RM$, 0 otherwise.
– $P_{c,\bar{c}}$ = Probability of students going directly from course section c to course section $\bar{c}$.

Proposed model for any given day:

$$\text{min:} \qquad \sum_{l,\bar{l}\in RM} \sum_{c,\bar{c}\in\overline{CRN}} P_{c,\bar{c}}N_{c,\bar{c}}(1-A_{l,\bar{l}})x_{c,l}x_{\bar{c},\bar{l}} \qquad (1)$$

$$\text{subject to:} \qquad \sum_{l} x_{c,l} = 1 \qquad\qquad \forall\, c \qquad (2)$$

$$\sum_{c\in O_t} x_{c,l} \leq 1 \qquad\qquad \forall\, l \in RM,\, t \in T \quad (3)$$

$$x_{c,l} \leq B_{c,l} \qquad\qquad \forall\, c \in CRN,\, l \in RM \qquad (4)$$

$$\text{and} \qquad x_{c,l} \in \{0,1\} \qquad\qquad \forall\, c \in CRN,\, l \in RM. \qquad (5)$$

where $P_{c,\bar{c}}$ is the uncertain parameter. Let $\mathcal{X}$ represent the feasible region obtained from Eqs. (2)–(5). Let $\mathcal{U} = \{P_{c,\bar{c}}|l_{c,\bar{c}} \leq P_{c,\bar{c}} \leq u_{c,\bar{c}}\}$ represent the uncertainty in estimating the probabilities for the students. The above formulation can be updated as:

$$\min_{x_{c,l}\in\mathcal{X}} \qquad \max_{P_{c,\bar{c}}\in\mathcal{U}} \sum_{l,\bar{l}\in RM} \sum_{c,\bar{c}\in\overline{CRN}} N_{c,\bar{c}}(1-A_{l,\bar{l}})P_{c,\bar{c}}x_{c,l}x_{\bar{c},\bar{l}} \qquad (6)$$

Clearly, Formulation (6) is nonlinear. However, it can be linearized efficiently. Let $y_{c,l,\bar{c},\bar{l}} \in [0,1]$ be defined as follows:

$$y_{c,l,\bar{c},\bar{l}} \leq x_{c,l} \qquad \forall c, l, \bar{c}, \bar{l} \tag{7a}$$

$$y_{c,l,\bar{c},\bar{l}} \leq x_{\bar{c},\bar{l}} \qquad \forall c, l, \bar{c}, \bar{l} \tag{7b}$$

$$y_{c,l,\bar{c},\bar{l}} \geq x_{c,l} + x_{\bar{c},\bar{l}} - 1 \qquad \forall c, l, \bar{c}, \bar{l} \tag{7c}$$

Using the above transformations, the formulation is updated as:

$$\min_{x_{c,l}, y_{c,l,\bar{c},\bar{l}}} : \quad \max_{P_{c,\bar{c}} \in \mathcal{U}} \sum_{l,\bar{l} \in RM} \sum_{c,\bar{c} \in \overline{CRN}} N_{c,\bar{c}}(1 - A_{l,\bar{l}}) \, P_{c,\bar{c}} \, y_{c,l,\bar{c},\bar{l}} \tag{8}$$

subject to:

$$(2)-(5), (7) \tag{9}$$

Notice that the inner problem is linear. Thus, taking the dual of the inner problem, we get the following:

$$\min_{x_{c,l}, y_{c,l,\bar{c},\bar{l}}} : \quad \min_{\alpha_{\bar{c}}, \beta_{\bar{c}}} \sum_{c,\bar{c} \in \overline{CRN}} (-l_{c,\bar{c}} \alpha_{c,\bar{c}} + u_{c,\bar{c}} \beta_{c,\bar{c}}) \tag{10}$$

subject to:

$$-\alpha_{c,\bar{c}} + \beta_{c,\bar{c}} \geq N_{c,\bar{c}} \sum_{l,\bar{l} \in RM} (1 - A_{l,\bar{l}}) \, y_{c,l,\bar{c},\bar{l}} \qquad \forall \bar{c} \in \overline{CRN} \tag{11}$$

$$\alpha_{\bar{c}}, \beta_{\bar{c}} \geq 0 \qquad \forall \bar{c} \in \overline{CRN} \tag{12}$$

$$(2)-(5), (7) \tag{13}$$

where $\alpha_{\bar{c}}$ and $\beta_{\bar{c}}$ are the dual variables corresponding to the constraints in $\mathcal{U}$. So, the robust linear model will be:

$$\min_{x_{c,l}, y_{c,l,\bar{c},\bar{l}}, \alpha_{\bar{c}}, \beta_{\bar{c}}} : \quad \sum_{c,\bar{c} \in \overline{CRN}} (-l_{c,\bar{c}} \alpha_{c,\bar{c}} + u_{c,\bar{c}} \beta_{c,\bar{c}}) \tag{14}$$

subject to:

$$(2)-(5), (7), (11)-(12). \tag{15}$$

Now, strengthening the formulation using RLT [12], we get the following additional constraints:

$$\sum_{l \in RM} y_{c,l,\bar{c},\bar{l}} = x_{\bar{c},\bar{l}} \qquad \forall \, c, \bar{c} \in CRN, \, \bar{l} \in RM \tag{16}$$

$$\sum_{c \in O_t} y_{c,l,\bar{c},\bar{l}} \leq x_{\bar{c},\bar{l}} \qquad \forall \, \bar{c} \in CRN, \, \bar{l} \in RM, \, t \in T \tag{17}$$

$$y_{c,l,\bar{c},\bar{l}} \leq B_{c,l} \qquad \forall \, c, \bar{c} \in CRN, \, l, \bar{l} \in RM \tag{18}$$

$$y_{c,l,\bar{c},\bar{l}} \leq B_{\bar{c},\bar{l}} \qquad \forall \, c, \bar{c} \in CRN, \, l, \bar{l} \in RM \tag{19}$$

To sum, the overall robust CAP model is:

$$\min_{x_{c,l}, y_{c,l,\bar{c},\bar{l}}, \alpha_{\bar{c}}, \beta_{\bar{c}}} \quad \sum_{c,\bar{c} \in \overline{CRN}} \left(-l_{c,\bar{c}} \alpha_{c,\bar{c}} + u_{c,\bar{c}} \beta_{c,\bar{c}} \right)$$

subject to:

$$(2)-(4), (7), (11)-(12), (16)-(17). \tag{20}$$

where Eqs. (5, 18, 19) are typically used during the pre-processing phase of solving the CAP model. In the next section, a numerical case study is illustrated to depict the performance of the proposed CAP model.

4 Numerical Study

In this section we present a numerical study carried out on synthetic scenario. The crux of the synthetic scenario is to generate a random but feasible course timetabling schedule. The key parameters of the scenario include the total number of students $n = 400$, and total number of course sections offered $C = 60$ in a given day. The duration of each course is set to one hour, and we assume that any section can be assigned to any classroom location. All the courses and students are divided into $R = 10$ groups, emulating R programs or majors. The number of students in each program is divided into two groups, one taking courses within their program only, and the other group that can take one or two courses from Programs-1, 2 or 3. The entire day is divided into 8 time slots. Every student will have 4 courses per day to take, and every department will offer 6 courses per day. The values of $l_{c,\bar{c}}$ and $u_{c,\bar{c}}$ are randomly generated such that $l_{c,\bar{c}} \leq u_{c,\bar{c}}$ for all $c, \bar{c} \in C$. The values of $A_{l,\bar{l}}$ for all $l, \bar{l} \in RM$ is presented in Fig. 3, where $|RM| = 12$, and the rows/columns indices represents the classroom labels. The dark colored cells in the figure indicate that the adjacent classrooms are at walkable distance. On the other hand, the white colored cells in the figure indicate that the students take buses to move between the corresponding classes. The total number of sections per time slot is depicted in Fig. 4, which indicates that the synthetic schedule is uniformly spread across all the time slots.

For the synthetic scenario, three strategies are used to assign sections to the classrooms. The first strategy is to assign the class location to the nearest available location to the program, say greedy strategy. The second strategy is to assign locations randomly, say random strategy. In random strategy, 30 replications are performed to identify possible spectrum of random solutions. Finally, we solve the proposed MIP to obtain the class room locations. The best solution obtained after 120 mins is shown in the following analysis. All of the proposed strategies are programmed in Python [13] on Windows 64 bit operating system, and the MIP is solved using Gurobi 9.5.2 solver [6].

Figure 5 presents a comparison of the greedy strategy with the MIP solution. In the figure, the red colored line depicts the MIP solution. The ticks on the x-axis represent distinct students, and the values on y-axis represent total number of times per day traveled in buses (or non-essential interactions). Similarly, Fig. 6

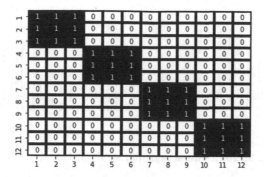

Fig. 3. Walk-ability between classrooms

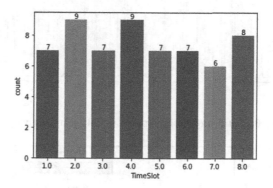

Fig. 4. Total sections per time slot

presents a comparison of 30 random assignments with the MIP solution. In both figures, we can see that MIP solution outperforms greedy and random assignments. The overall (total interactions per day of all the students) is depicted in Fig. 7, where the red colored line at the bottom of the plot indicates the total interactions in MIP solution, the black colored line on the left indicates the interactions via greedy approach, and box plot depicts interactions via random assignments. For the random assignment, the average total interactions are 782.5, for greedy strategy the total interactions are 227, and from MIP the total interactions are 207. To sum, MIP solution is roughly 10% better compared to greedy assignments, and 278% better compared to random assignments.

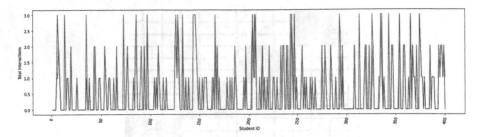

Fig. 5. Total interactions per student: Greedy vs MIP

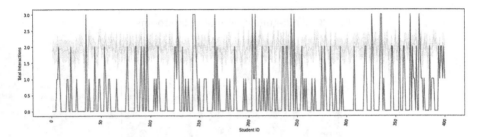

Fig. 6. Total interactions per student: Random vs MIP

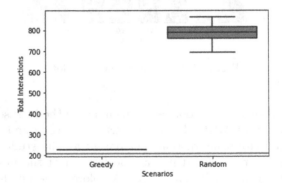

Fig. 7. Summary of overall level of interactions

5   Conclusion

A novel robust CAP that incorporates the interactions and behavior of the university level students is presented in the paper. The objective of the proposed CAP model is to minimize the total non-essential interactions among the university level students. An initial quadratic measure that estimates total interactions is developed. Then the model is reformulated and the final version of the model boils down to a linear MIP. The toy numerical case study indicates that the proposed approach reduces the overall interactions among the students. Although the numerical case study is synthetic, it portrays the applicability of the proposed approach to the real-world CAP scenarios. In this work, the behavior of the students is captured through the probabilities, and the robustness is incorporated through the bounds on the probabilities. The uncertainty depicted via bounds can be estimated through data collections and/or surveys in real-world scenarios.

Acknowledgments. The work is funded by the Interdisciplinary Research Center for Intelligent Secure Systems at King Fahd University of Petroleum & Minerals (KFUPM) under the grant INSS2208.

References

1. Burkard, R.E., Cela, E., Pardalos, P.M., Pitsoulis, L.S.: The Quadratic Assignment Problem. Springer, Heidelberg (1998). https://doi.org/10.1007/978-1-4757-2787-6
2. Burke, E., Elliman, D., Ford, P., Weare, R.: Examination timetabling in British universities: a survey. In: Burke, E., Ross, P. (eds.) PATAT 1995. LNCS, vol. 1153, pp. 76–90. Springer, Heidelberg (1996). https://doi.org/10.1007/3-540-61794-9_52
3. Burke, E.K., Petrovic, S.: Recent research directions in automated timetabling. Eur. J. Oper. Res. **140**(2), 266–280 (2002)
4. Carter, M.W., Tovey, C.A.: When is the classroom assignment problem hard? Oper. Res. **40**(1–supplement–1), S28–S39 (1992)
5. Elloumi, A., Kamoun, H., Jarboui, B., Dammak, A.: The classroom assignment problem: complexity, size reduction and heuristics. Appl. Soft Comput. **14**, 677–686 (2014)
6. Gurobi Optimization, LLC: Gurobi Optimizer Reference Manual (2023). https://www.gurobi.com
7. Koopmans, T.C., Beckmann, M.: Assignment problems and the location of economic activities. Econom.: J. Econ. Soc. 53–76 (1957)
8. Kuhn, H.W.: The Hungarian method for the assignment problem. Naval Res. Logist. Q. **2**(1–2), 83–97 (1955)
9. Lawler, E.L.: The quadratic assignment problem. Manage. Sci. **9**(4), 586–599 (1963)
10. Motzkin, T.: The assignment problem[1]. Numer. Anal. (6), 109 (1956)
11. Phillips, A.E., Waterer, H., Ehrgott, M., Ryan, D.M.: Integer programming methods for large-scale practical classroom assignment problems. Comput. Oper. Res. **53**, 42–53 (2015)
12. Sherali, H.D., Adams, W.P.: A Reformulation-Linearization Technique for Solving Discrete and Continuous Nonconvex Problems, vol. 31. Springer, Heidelberg (2013)
13. Van Rossum, G., Drake Jr., F.L.: Python reference manual. Centrum voor Wiskunde en Informatica Amsterdam (1995)
14. Wren, A.: Scheduling, timetabling and rostering—a special relationship? In: Burke, E., Ross, P. (eds.) PATAT 1995. LNCS, vol. 1153, pp. 46–75. Springer, Heidelberg (1996). https://doi.org/10.1007/3-540-61794-9_51

Introducing a New Metric for Improving Trustworthiness in Real Time Object Detection

Konstantinos Tarkasis[✉][iD], Konstantinos Kaparis[iD],
and Andreas C. Georgiou[iD]

Quantitative Methods and Decision Analysis Lab, Department of Business
Administration, University of Macedonia, 156 Egnatia St, 54636 Thessaloniki, Greece
{ktarkasis,k.kaparis,acg}@uom.edu.gr

Abstract. This preliminary study, investigates the possible benefits of
using a novel metric in real-time evaluation of outputs generated by Real-
Time Object Detection Algorithms. Our primary goal is to improve the
reliability of the detection process through the spatial analysis of the
sequence of video frames used as input data for a Convolutional Neural
Network (CNN). The method focuses on the analysis of the variations
between consecutive output values from the CNN. By leveraging established similarity metrics, we try to identify patterns that signal potential instances of false positive predictions and develop a methodology-
agnostic assessment of the CNN's output quality. The paper concludes
with some preliminary computational results that support the efficacy
and potential applications of the proposed method.

Keywords: Real Time Object Detection · Trustworthiness · xAI ·
CNN

1 Introduction

This work investigates a new algorithmic scheme, that builds upon existing state-
of-the-art models on real time object detection [2,4–7,13,15–18,21,23].

This investigation evaluates the "trust" of a CNN by detecting "anomalies"
and acts as a second layer of consistency. For our purposes, we use YOLO v4
[2] and YOLO v7 [21]. However, any computer vision object detection scheme
which upon call, returns the coordinates of the bounding boxes along with the
confidence score for the attempted classification, could be applied as well.

2 Literature Review

Although there exists a considerable body of literature on object tracking in consecutive frames, our interest is more relevant to papers like, Sternig *et al.* [20] who
harnessed full geometry and the object's center of mass to create an efficient 3D

H. Moosaei et al. (Eds.): DIS 2023, LNCS 14321, pp. 242–249, 2024.
https://doi.org/10.1007/978-3-031-50320-7_18

object tracking algorithm. Also, Kalal *et al.* [10] introduced a framework related to tracking, learning, and detection assuming limited frame-to-frame motion and object visibility. Bewley *et al.* [1] investigated the association of detections with existing targets.

Yoshihashi *et al.* [24] introduced the "search window strategy" and suggested limiting search windows based on the object's physical speed. Multitarget tracking [14] and region-based detection methods [28] involve object identification via bounding boxes, followed by feature extraction for track association. Zhou *et al.* [27] use IoU-based matching to link objects across frames. Lezki *et al.* [11] also incorporated object center tracking with sparse optical flow. Lastly, Wang *et al.* [22] utilize multiple consecutive frames to predict item trajectories by calculating trajectories fitted by continuous frames based on bounding box data.

In the last decade, significant developments have been made towards algorithms focusing on processing live video streams and static images. In this respect, the reduction of false positives remains significant and mainly involves optimizing neural network architectures.

In this respect, RCNNs (Region-based Convolutional Neural Network) [7], employed a two-stage process involving region proposal and object classification and Fast RCNNs, while [6] eliminated the region proposal step, by directly feeding input to the CNN. Faster RCNNs [18] improved efficiency by integrating region proposal networks and removing the selective search algorithm. Also, Single Shot MultiBox Detector (SSD) [13] introduced a one-stage architecture for faster detection. On the other hand, YOLOv2 [16] and YOLOv3 [17] refined bounding box predictions with anchor boxes, YOLOv4 [2] added further optimizations like PANet and CIoU [26]. Finally, YOLOv7 [21] integrated the extended Efficient Layer Aggregation Network (E-ELAN).

All these efforts represent ongoing research in creating neural networks capable of handling diverse computer vision tasks beyond mere object tracking. This work focuses on real-time motion object detection in consecutive video frames within this subdomain and it is a preliminary attempt in employing a novel metric called the Sequential Intersection over Union (sIoU).

3   Methodology

3.1   The Sequential Intersection over Union

The increase of computational power over the past few years, allows us to introduce a modified perspective of the so-called *Intersection Over Union* (IoU) metric, based on the *Jaccard similarity coefficient* [9]. In particular, as we know, the IoU is used to evaluate a model's ability to distinguish objects from their backgrounds in images [3,25]. Moreover, it measures the *localization accuracy* and *localization errors* of such models by calculating the overlap between the predicted bounding box and the ground truth bounding box. This is expressed as the following ratio

$$IoU = \frac{\text{Area of Overlap}}{\text{Area of Union}}.$$

We use this concept to expand the evaluation capability of the employed algorithm, by calculating the IoU of two consecutive frames. In this respect we denote F_t and F_{t+1} the coordinates of the bounding boxes for two consecutive frames in time t and $t+1$ respectively, we define the *sequential Intersection over Union* (sIoU) in $t+1$ as the following ratio

$$\text{sIoU}_{t+1} = \frac{F_t \cap F_{t+1}}{F_t \cup F_{t+1}}.$$

In such a process we distinguish the following four cases:

1. sIoU = Nan: Both bounding boxes are empty ($F_t = F_{t+1} = \emptyset$)
2. $0 < \text{sIoU} < 1$: The intersection of the bounding boxes is not empty ($F_t \cap F_{t+1} \neq \emptyset$).
3. sIoU $= 0$: The intersection of the bounding boxes is empty ($F_t \cap F_{t+1} = \emptyset$ and $F_t \cup F_{t+1} \neq \emptyset$).
4. sIoU $= 1$: Both bounding boxes coincide and thus the item under consideration has not moved during the transition from t to $t+1$ ($F_t = F_{t+1} \neq \emptyset$).

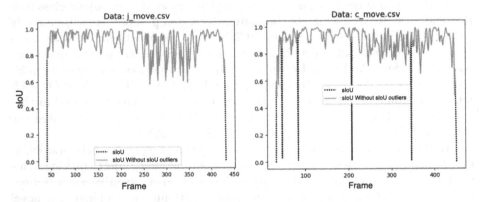

Fig. 1. sIoU over Frames for videos without (j_movie) and with (c_movie) false positive instances.

By the definition of sIoU, we expect values close to 1 assuming that we are considering a true positive item.

An item should be at almost the same place in the transition from t to $t+1$ when we consider true positive observations. On the other hand, false positive observations emerge sporadically and their sIoU should be very low or NaN (Fig. 1). Our evaluation is done upon those sIoU scores and attempts to identify frames which contain noise and outliers.

3.2   A False Positive Filtering Procedure

Through the sIoU metric, we have at our disposal the means to identify potential false positive observations. An indicative procedure to exploit this metric is the following:

Step 1 Use the vector of *binary values* denoting the presence of the pre-defined item within the confines of a specific bounding box, and the accompanying *confidence scores* and *bounding boxes* coordinates to identify data frames that potentially contain false positive observations.

Step 2 Impute false positives [12] using the mean of the preceding and succeeding frames' confidence scores.

In this respect we modify the confidence score of identified objects that we consider having been inaccurately detected, replacing it with the average from frames that we deem reliable.

Step 3 Perform polynomial interpolation of the confidence score that best fits to the imputed dataset.

Step 4 Employ the aforementioned polynomial to compute the residual errors of the confidence score values generated by the neural network in the second step of the process.

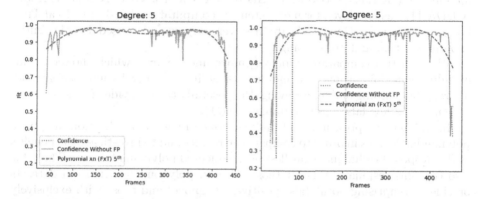

Fig. 2. Polynomial function of degree 5 for the approximation of confidence score.

In the case under investigation, through parametric and/or empirical tests (e.g. DBSCAN [19]) the threshold for discerning false positive identifications was set at a value below 0.4. Also, based on the evidence provided by investigating polynomial degrees ranging from 3 to 9, we employed a degree of five (5) (see Fig. 2). In addition, through experimentation, we concluded that a residual error threshold indicative of a video without false positives, resides below 0.9. Conversely, for instances where the video encompasses false positives, the value increases beyond the threshold of 1.5.

When the residual error of a neural network returns a value below the predetermined threshold (i.e. <0.9), it is highly likely that the video does not contain false positives. Conversely, a higher residual error signifies instances where the neural network has indeed identified false positives, but with a subdued confidence score. This observation, though seemingly desirable due to the low confidence score, is by no means the performance of an accurate neural network.

Instead, it prompts a local search of frames highlighted by the confidence score and sIoU.

4   Computational Experiments

Data processing tasks were performed using Python (versions 3.9 and 3.10) whereas, for object detection and recognition tasks, YoloV7 [21] and YoloV4 [2] architectures were employed, utilizing publicly available pre-trained weights[1,2], with pro-evaluate accuracy of the models.

In terms of datasets, the study utilized the i3dpost multi-view and 3D human action/interaction video dataset, which originates from the collaborative efforts of the University of Surrey in the UK and CERTH-ITI in Greece [8]. Additionally, custom datasets were tailored to meet specific research requirements, ensuring a comprehensive dataset portfolio.

The experiments were carried out on a PC running under Windows 11 with an Intel i7 (Gen11) processor, 16 GB of RAM and a GPU Nvidia RTX3070 (CUDA 11, was employed for more intensive computational tasks like Real Time Object Detection). Google Colab (free version) was also used for the initial experimentation and prototyping.

We used two comparable videos (j_move and c_move) which showed a pair of individuals doing some slow movements against a natural white background, followed by jumping and limb waving. Purposefully the first video (c_move) tends to generate a few false positive frames (Fig. 1).

In Table 1, we present initial findings concerning the application of typical polynomial degrees in order to compare the results and to investigate the benefits of the proposed technique regardless of the selected polynomial degree. The table also contains residual errors for those selected polynomial degrees on datasets or videos comprising both "false positives" (c_move) and those with exclusively "true positives" (all videos excluding c_move). The row labeled "sIoU num" provides the count of sub-threshold sIoU detections.

The results show that the first video (j_move) has very low residual error values, while the second video (c_move) has values close to 2.0. Additionally, as the polynomial degree increases, we observe a gradual decrease in residual errors, highlighting the trade-off between polynomial complexity and reduced residual errors.

Additional experiments conducted using video data generated under controlled laboratory conditions [8], demonstrate minimal residual error values due to the close correspondence between the fitted function and the actual data. Notably, the videos categorized as 'D' in Table 1 depict a human subject transitioning across the frame with variations in entry and exit angles and directions. Despite the rapid movement of the subject in these videos, the findings remain consistent, primarily owing to the recording's frame rate exceeding the subject's velocity requirements.

---

[1] https://github.com/WongKinYiu/yolov7.git.
[2] https://github.com/AlexeyAB/darknet.git.

In conclusion, this study initially posited the emergence of discernible patterns when a human or object moves between successive frames, and these experiments have indeed validated the hypothesis, yielding a cohesive outcome.

Table 1. Residual error results

Data	j_move	c_move	D1-03_03	D1-29_00	D1-03_07	D1-16_03
sIoU num	2	10	2	4	3	2
Degree	Residual Error					
3	0,704	3,39	0,479	0,804	0,528	0,479
5	**0,383**	**2,459**	0,086	0,454	0,468	0,307
9	0,169	1,963	0,249	0,016	0,173	0,075
Data	D1-29_04	D1-42_02	D2-15_00	D2-15_05	D2-30_05	D2-42_02
sIoU num	1	3	1	1	2	2
Degree	Residual Error					
3	0,08	0,085	0,266	0,182	0,311	0,218
5	0,04	0,056	0,158	0,042	0,183	0,183
9	0,027	0,007	0,014	0,016	0,129	0,021

5   Limitations and Further Research

In this preliminary inquiry, our emphasis was on the introduction of a novel metric and its investigative incorporation into a sequential procedure for identification and categorization of single entities within individual frames. Subsequent endeavors are required to address the requirements for multiple categories, such as the pre-filtering of entities based on their respective types prior to the implementation of the proposed methodology.

In addition, due to the utilization of bounding boxes, it is crucial to synchronize the frame rate with the velocity of the specific object and attaining such elevated frame rates might not be viable in real-time applications.

Another known issue that occurs during the training of neural networks is the tendency to favor a specific object category, leading to inflated confidence scores even for incorrect identifications. While such cases often yield minor residual errors, there are occasions where the sIoU value drops below 0.4, indicating false positive results. One potential approach to address this issue is the inclusion of a regularization function combining residual errors from the original dataset with errors arising during coefficient computation for the fitted function. Our current research focuses in determining the appropriate coefficients and weights for this function to align with the existing framework under investigation.

Acknowledgement. This work is part of a project that has received funding from the Research Committee of the University of Macedonia under the Enhancing Research 2023 funding programme.

References

1. Bewley, A., Ge, Z., Ott, L., Ramos, F., Upcroft, B.: Simple online and realtime tracking. In: 2016 IEEE International Conference on Image Processing (ICIP), pp. 3464–3468. IEEE (2016)
2. Bochkovskiy, A., Wang, C.Y., Liao, H.Y.M.: YOLOv4: optimal speed and accuracy of object detection. arXiv preprint arXiv:2004.10934 (2020)
3. Böttger, T., Follmann, P., Fauser, M.: Measuring the accuracy of object detectors and trackers. In: Roth, V., Vetter, T. (eds.) GCPR 2017. LNCS, vol. 10496, pp. 415–426. Springer, Cham (2017). https://doi.org/10.1007/978-3-319-66709-6_33
4. Broad, A., Jones, M., Lee, T.Y.: Recurrent multi-frame single shot detector for video object detection. In: BMVC, p. 94 (2018)
5. Dai, J., Li, Y., He, K., Sun, J.: R-FCN: object detection via region-based fully convolutional networks. In: Advances in Neural Information Processing Systems, pp. 379–387 (2016)
6. Girshick, R.: Fast R-CNN. In: Proceedings of the IEEE International Conference on Computer Vision, pp. 1440–1448 (2015)
7. Girshick, R., Donahue, J., Darrell, T., Malik, J.: Rich feature hierarchies for accurate object detection and semantic segmentation. In: Proceedings of the IEEE Conference on Computer Vision and Pattern Recognition, pp. 580–587 (2014)
8. Gkalelis, N., Kim, H., Hilton, A., Nikolaidis, N., Pitas, I.: The I3Dpost multi-view and 3D human action/interaction database. In: 2009 Conference for Visual Media Production, pp. 159–168. IEEE (2009)
9. Jaccard, P.: The distribution of the flora in the alpine zone. 1. New Phytol. **11**(2), 37–50 (1912)
10. Kalal, Z., Mikolajczyk, K., Matas, J.: Tracking-learning-detection. IEEE Trans. Pattern Anal. Mach. Intell. **34**(7), 1409–1422 (2011)
11. Lezki, H., et al.: Joint exploitation of features and optical flow for real-time moving object detection on drones. In: Proceedings of the European Conference on Computer Vision (ECCV) Workshops (2018)
12. Little, R.J., Rubin, D.B.: Statistical Analysis with Missing Data, vol. 793. Wiley, Hoboken (2019)
13. Liu, W., et al.: SSD: single shot multibox detector. In: Leibe, B., Matas, J., Sebe, N., Welling, M. (eds.) ECCV 2016. LNCS, vol. 9905, pp. 21–37. Springer, Cham (2016). https://doi.org/10.1007/978-3-319-46448-0_2
14. Mahmoudi, N., Ahadi, S.M., Rahmati, M.: Multi-target tracking using CNN-based features: CNNMTT. Multimed. Tools Appl. **78**, 7077–7096 (2019)
15. Redmon, J., Divvala, S., Girshick, R., Farhadi, A.: You only look once: unified, real-time object detection. In: Proceedings of the IEEE Conference on Computer Vision and Pattern Recognition, pp. 779–788 (2016)
16. Redmon, J., Farhadi, A.: YOLO9000: better, faster, stronger. In: Proceedings of the IEEE Conference on Computer Vision and Pattern Recognition, pp. 7263–7271 (2017)
17. Redmon, J., Farhadi, A.: YOLOv3: an incremental improvement. arXiv preprint arXiv:1804.02767 (2018)
18. Ren, S., He, K., Girshick, R., Sun, J.: Faster R-CNN: towards real-time object detection with region proposal networks. Adv. Neural. Inf. Process. Syst. **28**, 91–99 (2015)
19. Sander, J., Ester, M., Kriegel, H.P., Xu, X.: Density-based clustering in spatial databases: the algorithm GDBSCAN and its applications. Data Min. Knowl. Disc. **2**(2), 169–194 (1998)

20. Sternig, S., Riemenschneider, H., Roth, P.M., Donoser, M., Bischof, H.: Robust person detection by classifier cubes and local verification (2010)
21. Wang, C.Y., Bochkovskiy, A., Liao, H.Y.M.: YOLOv7: trainable bag-of-freebies sets new state-of-the-art for real-time object detectors. arXiv preprint arXiv:2207.02696 (2022)
22. Wang, Y., Cheng, H., Zhou, X., Luo, W., Zhang, H.: Moving ship detection and movement prediction in remote sensing videos. Int. Arch. Photogram. Remote Sens. Spat. Inf. Sci. **43**, 1303–1308 (2020)
23. Wu, B., Iandola, F., Jin, P.H., Keutzer, K.: SqueezeDet: unified, small, low power fully convolutional neural networks for real-time object detection for autonomous driving. In: Proceedings of the IEEE Conference on Computer Vision and Pattern Recognition Workshops, pp. 129–137 (2017)
24. Yoshihashi, R., Trinh, T.T., Kawakami, R., You, S., Iida, M., Naemura, T.: Learning multi-frame visual representation for joint detection and tracking of small objects (2017)
25. Yu, J., Jiang, Y., Wang, Z., Cao, Z., Huang, T.: UnitBox: an advanced object detection network. In: Proceedings of the 24th ACM International Conference on Multimedia, pp. 516–520 (2016)
26. Zheng, Z., Wang, P., Liu, W., Li, J., Ye, R., Ren, D.: Distance-IoU loss: faster and better learning for bounding box regression. In: Proceedings of the AAAI Conference on Artificial Intelligence, vol. 34, pp. 12993–13000 (2020)
27. Zhou, X., Koltun, V., Krähenbühl, P.: Tracking objects as points. In: Vedaldi, A., Bischof, H., Brox, T., Frahm, J.-M. (eds.) ECCV 2020. LNCS, vol. 12349, pp. 474–490. Springer, Cham (2020). https://doi.org/10.1007/978-3-030-58548-8_28
28. Zhou, X., Zhuo, J., Krahenbuhl, P.: Bottom-up object detection by grouping extreme and center points. In: Proceedings of the IEEE/CVF Conference on Computer Vision and Pattern Recognition, pp. 850–859 (2019)

Author Index

© The Editor(s) (if applicable) and The Author(s), under exclusive license
to Springer Nature Switzerland AG 2024
H. Moosaei et al. (Eds.): DIS 2023, LNCS 14321, pp. 251–252, 2024.
https://doi.org/10.1007/978-3-031-50320-7

Printed in the United States
by Baker & Taylor Publisher Services